MATHEMATICS RESEARCH DEVELOPMENTS

MATHEMATICAL THEORIES, MODELS AND GROUPS

Mathematics Research Developments

Additional books in this series can be found on Nova's website under the Series tab.

Additional E-books in this series can be found on Nova's website under the E-books tab.

MATHEMATICS RESEARCH DEVELOPMENTS

MATHEMATICAL THEORIES, MODELS AND GROUPS

MARCUS J. GRECHO
EDITOR

Nova Science Publishers, Inc.
New York

For permission to use material from this book please contact us:
Telephone 631-231-7269; Fax 631-231-8175
Web Site: http://www.novapublishers.com

LIBRARY OF CONGRESS CATALOGING-IN-PUBLICATION DATA

Mathematical theories, models, and groups / [edited by] Marcus J. Grecho.
p. cm.
Includes index.
ISBN 978-1-61324-250-6 (hardcover)
1. Game theory. 2. Group theory. I. Grecho, Marcus J.
QA269.M386 2011
519.3--dc22
2011009068

Published by Nova Science Publishers, Inc. † New York

CONTENTS

PREFACE

This new book presents current research in the field of mathematics, with a particular emphasis on mathematical theories, models and groups. Topics discussed include particular swarm optimization in theory and practice; detecting the commuting probability of the derived subgroup; peer group situations and games with interval uncertainty; groups of wraps and diffeomorphisms of manifolds over non-archimedean normed fields and the concept of a suitable insurance policy using leader-follower games.

A research area, which has gained increasing attention in recent years, investigates the application of biological concepts to various optimization tasks in science and technology. Technically, global extrema of objective functions in a *d*-dimensional discrete or continuous space have to be determined or approximated, which is a standard problem with an ample number of applications. In Chapter 1 the particle swarm optimization paradigm is described in different variations of the equations of motion and studied comparatively in theory and empirically on the basis of selected optimization problems. To facilitate this task, the different variants are described in a general scheme for optimization algorithms. Then different side constraints as initial conditions, boundary conditions of the search space and velocity restrictions are investigated in detail.

This is followed by an efficiency comparison of swarm optimization to a selection of other global optimization heuristics, such as various single state iterative search methods (simulated annealing, threshold accepting, great deluge algorithm, basin hopping, etc.), ensemble based algorithms (ensemble-based simulated annealing and threshold accepting), and evolutionary approaches. In specific, the application of these algorithms to combinatorial problems and standard benchmark functions in high-dimensional search spaces is examined. Further, the authors show how particle swarm optimization can be effectively used for the optimization of so-called single-warehouse multi-retailer systems to optimize the ordering strategy and the transportation resources. Optimization tasks of this class are very common in economy as e.g. in the manufacturing industry and for package delivery services.

There is a long standing line of research, which is devoted to investigate bounds for ♣G' ♣when G is an infinite group. This line goes back to a classic result of I. Schur. Chapter 2 deals with the structure of G' when G is a compact group, showing that ♣G' ♣ can be controlled by the notion of commuting probability.

Chapter 3 presents an historical account on higher derivations, generalized higher derivations and other kinds of higher derivations in rings, based on the work of several authors. Moreover, recent results as well as some possible directions for future researches on these topics have been discussed in details. Finally, some examples and applications of higher derivations have been given.

In Chapter 4, by using the introduced notions, the authors approach the study of peer group games in an interval-valued setting. Peer group games have been introduced in and correspond to situations where the players have a hierarchical relationship and a positive real value is given to each player only if all his superiors cooperate with him. Peer group games are a subclass of the class of trading games and also of the class of games with permission structure. The authors introduce interval peer group situations and for the related interval peer group games the authors give properties of monotonicity, convexity and specify the Shapley value and the Weber set. In the last section all the results are applied to sealed bid second price auctions with uncertainty.

In Chapter 5 by an invariance-property of solutions for shifted games, the authors define a subset of the reasonable set of outcomes and characterize it. This set includes the prekernel and the least core, so it has a non-empty intersection with the core when the core is non-empty. Furthermore the authors characterize the set of pairwise reasonable outcomes.

Chapter 6 is devoted to the investigation of groups of diffeomorphisms and wraps of manifolds over non-archimedean fields of zero and positive characteristics. Different types of topologies are considered on groups of wraps and diffeomorphisms relative to which they are generalized Lie groups or topological groups. Among such topologies pairwise incomparable are found as well. Topological perfectness of the diffeomorphism group relative to certain topologies is studied. There are proved theorems about projective limit decompositions of these groups and their compactifications for compact manifolds. Moreover, an existence of one-parameter local subgroups of diffeomorphism groups is investigated.

Chapter 7 focuses on a new method to create a suitable new insurance product; the aim is, obviously, to satisfy the needs of the potential policy holders maximizing the expected utility both of the company and the agent, as far as possible. By analyzing the Italian life insurance market, the authors can observe a new trend for the companies that aim to create policies for different sets of policy owners. Life policy can today be custom-made to suit the client's needs and new unit-linked policies have been conceived precisely to reach a specific segment of the market.

Such a behavior represents a new approach for the companies, which so far have not focused their attention on this problem. This explains why no relevant actuarial literature on this topic exists.

In: Mathematical Theories, Models and Groups ISBN 978-1-61324-250-6
Editor: Marcus J. Grecho

Chapter 1

Comparative Study of Different Approaches to Particle Swarm Optimization in Theory and Practice

***Stefanie Thiem*[1,2] *and Jörg Lässig*[3]**
[1]Institut für Physik, Technische Universität Chemnitz, Chemnitz, Germany
[2]Department of Mathematics and Statistics, The Open University, Milton Keynes, UK
[3]International Computer Science Institute, Berkeley, CA, US

Abstract

A research area, which has gained increasing attention in recent years, investigates the application of biological concepts to various optimization tasks in science and technology. Technically, global extrema of objective functions in a d-dimensional discrete or continuous space have to be determined or approximated, which is a standard problem with an ample number of applications. In this chapter the particle swarm optimization paradigm is described in different variations of the equations of motion and studied comparatively in theory and empirically on the basis of selected optimization problems. To facilitate this task, the different variants are described in a general scheme for optimization algorithms. Then different side constraints as initial conditions, boundary conditions of the search space and velocity restrictions are investigated in detail.

This is followed by an efficiency comparison of swarm optimization to a selection of other global optimization heuristics, such as various single state iterative search methods (simulated annealing, threshold accepting, great deluge algorithm, basin hopping, etc.), ensemble based algorithms (ensemble-based simulated annealing and threshold accepting), and evolutionary approaches. In specific, the application of these algorithms to combinatorial problems and standard benchmark functions in high-dimensional search spaces is examined. Further, we show how particle swarm optimization can be effectively used for the optimization of so-called single-warehouse multi-retailer systems to optimize the ordering strategy and the transportation resources. Optimization tasks of this class are very common in economy as e.g. in the manufacturing industry and for package delivery services.

1. Introduction

In recent years several biological motivated optimization algorithms as *Genetic Algorithms* [1], *Artificial Immune Systems* [2] and *Ant Colony Optimization* [3] have been proposed.

Another possibility is the usage of the dynamics of swarms to find solutions to optimization problems. This algorithm was invented by Kennedy and Eberhart in 1995 and is referred to as *Particle Swarm Optimization* (PSO) [4]. The basic idea of PSO is the application of swarm behavior to different optimization problems. Originally, the algorithm was only introduced for problems with a continuous solution space but, meanwhile, also the application to combinatorial problems is possible [5, 6]. However, the latter is not considered in this chapter.

In addition, many different versions and additional heuristics have been introduced during the last years. In this chapter we focus on four of them. After introducing the algorithms, further heuristics and different boundary conditions are discussed in the next section. Then in Sec. 3. a performance comparison is accomplished based on a set of continuous benchmark functions by considering various aspects. Besides the overall performance of the different methods, we study the convergence behavior and the scaling behavior with increasing dimensionality and population size. Different aspects are also visualized by plots.

After comparing the different PSO variants, in Sec. 4. also other optimization heuristics are taken into account to compare the performance of the PSO approaches to them. More specific *Simulated Annealing* [7], *Threshold Accepting* [8], their *ensemble-based variants* [9], *Basin Hopping* [10] and *Evolution*

Strategies [11] are taken into account. They are combined with different move classes in the continuous domain, based on different distribution functions for the generation of new solutions.

2. The Particle Swarm Optimization Approach

The basic idea of the PSO algorithm is that a swarm of individuals can share information about the global so far best solution and additionally each individual has an internal memory to store its best so far solution. The movement of each particle is then given by a tradeoff between its current velocity (velocity component), a movement in the direction of its local best solution (cognitive component) and the global optimum (social component).

2.1. The Algorithm

To compare different variants of this approach later on we introduce a special notation. States will be specified by Greek letters α, β, etc. Since the algorithm works with more than one individual, the states are distinguished by subscripts, i.e. we write α_i, and all states together are written as a vector in bold letters $\boldsymbol{\alpha}$. Bringing this notation and the previously described ideas together we obtain a standard pattern as shown in Algorithm 1.

Algorithm 1 Particle Swarm Optimization

Input: problem $\mathcal{P}$ and a solution space $\mathcal{S}$
Output: solution $\gamma \in \mathcal{S}$

$\boldsymbol{\alpha} \leftarrow$ *compute_initial_states*$(a_1, a_2, \ldots, a_m)$
$\boldsymbol{\beta} \leftarrow \boldsymbol{\alpha}$
for all particles i **do**
    if ($\gamma =$ *null* or $f(\beta_i) < f(\gamma)$) **then**
        $\gamma \leftarrow \beta_i$
$\boldsymbol{\delta} \leftarrow \mathbf{0}$
repeat
    $\boldsymbol{\alpha}, \boldsymbol{\delta} \leftarrow$ *compute_new_state*$(\boldsymbol{\alpha}, \boldsymbol{\beta}, \gamma, \boldsymbol{\delta}, w, c_1, c_2)$
    for all particles i **do**
        if ($f(\alpha_i) < f(\beta)$) **then**
            $\beta_i \leftarrow \alpha_i$
            if ($f(\beta_i) < f(\gamma)$) **then**
                $\gamma \leftarrow \beta_i$
until (*termination_condition*$(d_1, d_2, \ldots, d_p)$)
return γ

The single parts of the algorithm are now explained in more detail. We begin with the different variables in the algorithm.

- The current state vector for all particles is α. In PSO the single state α_i is equivalent to the position $\mathbf{r}_i$ of the individual i in the solution space.
- The vector β is the vector of the local best states of each particle, i.e. the component β_i is equivalent to the best so far position $\mathbf{l}_i$ of individual i.
- The global best state is given by γ, which is equivalent to the best solution $\mathbf{g}$ of the complete swarm during the optimization run.
- The vector δ is not a solution but contains the current velocity vector for each particle. This vector can have very different forms depending on the optimization problem.
- The algorithm parameters w, c_1 and c_2 are user defined and can influence the performance of the algorithm. The parameter c_1 is the cognitive parameter and weights the influence of the local solution β_i in the construction of a new solution. In the same way the social parameter c_2 weights the influence of the so far best solution of the swarm γ. The weight factor w is responsible for the influence of the velocity vector.

We will briefly describe the three different functions included in this algorithm.

- The method *compute_initial_states*$(a_1, a_2, \ldots, a_m)$ computes an initial state for each particle i. It can depend on an arbitrary number of variables $a_1, a_2, \ldots, a_m$.
- The method *compute_new_state*$(\alpha, \beta, \gamma, \delta, w, c_1, c_2)$ is a method to compute a new state and velocity for each of the particles i. It depends on the parameters $\alpha, \beta, \gamma, \delta$, w, c_1, c_2 and there are many different internal realizations possible. The details for several realizations for the continuous domain are given in Sec. 2.2.1. to 2.2.4..
- The function *termination_condition*$(d_1, d_2, \ldots, d_p)$ returns true when the convergence criteria are fulfilled and can depend on an arbitrary number of variables $d_1, d_2, \ldots, d_p$.

Regarding the convergence criteria we have several options briefly discussed here:

- *Maximum number of evaluations:* With this option we perform the algorithm until a user defined number of objective function evaluations is reached.
- *Sufficient good approximation of the optimum:* In this case the loop is executed until the algorithm has found a solution $f(\gamma)$ which differs from the problem's best solution f_{best} by less than a given percentage, i.e. $|f(\gamma) - f_{\text{best}}| < \varepsilon$.[1]
- *No improvement in the last iterations:* The algorithm terminates when it has not found any improvements during the last iterations. The number of iterations can be specified by the user.
- *Contraction of the swarm:* This option is a special option only for swarm optimization, whereas the three others are also applicable for other algorithms. With this option we will stop the algorithm when the mean displacement of the swarm is reduced to a user defined value. This is possible because the swarm contracts itself near the so far global best solution.

2.2. Move Classes for Particle Swarm Optimization

So far we have not discussed how new states are constructed by the method *compute_ new_states*$(\alpha, \beta, \gamma, \delta, w, c_1, c_2)$. The original version of the PSO algorithm, which was presented by Kennedy and Eberhart in 1995 [4], realized it as shown in Listing 1. The matrix vx is aquivalent to our vector δ, where δ_i represents the velocity $\mathbf{v}_i$ of the individual i. The matrix `presentx` represents the currents state α, `pbest` the local best states β and the vector `gbest` is given by γ describing the overall best so far discovered state.

Listing 1: Original velocity and position update rule

```
vx[][] = vx[][] + 2*rand()*(pbest[][] - presentx[][])
                + 2*rand()*(gbest[] - presentx[][])
```

Later on scaling factors c_1 and c_2 replaced the constant factors 2 and Shi and Eberhart introduced a weighting factor for the velocity, receiving better results than for the original version [12]. Because the meaning of the term `rand()` is not unique, several different variants were proposed during the years

[1]Using the option of sufficient approximation, we have to know the exact value of the global optimum, which is not given for real world problems. Hence, this option is only useful for performance analysis tasks.

or even new suggestions for the update rules were introduced. Thus, in the following section we will present four different versions of PSO and compare their performance.

The first version interprets the equations given by Kennedy and Eberhart [4] in Listing 1 in the way that the `rand()` expression equals two scalar random numbers r_1 and r_2 which are multiplied with each vector component (see Sec. 2.2.1.). The other interpretation would be to think of random diagonal matrices $\mathcal{R}_1$ and $\mathcal{R}_2$ which weight each component of the velocity vectors differently (see Sec. 2.2.2.). Both methods have been applied in the past: for instance the first one was used in [13, 14] and the latter one in [4, 15].

2.2.1. Variant 1

The first version, denoted as PSO1, uses the velocity and position update rule

$$\mathbf{v}_i^{t+1} = w \cdot \mathbf{v}_i^t + c_1 \cdot r_1 \cdot \left(\mathbf{l}_i^t - \mathbf{r}_i^t\right) + c_2 \cdot r_2 \cdot \left(\mathbf{g}^t - \mathbf{r}_i^t\right) \tag{1a}$$

$$\mathbf{r}_i^{t+1} = \mathbf{r}_i^t + \mathbf{v}_i^t \quad , \tag{1b}$$

where r_1 and r_2 are uniform random numbers in $[0,1)$ which weight the importance of the global best solution $\mathbf{g}$ to the local best solution $\mathbf{l}_i$ of particle i on the way of finding a new position $\mathbf{r}_i^t$.

The algorithmic realization is quite simple and shown in Algorithm 2. The position and the velocity components for the different particles i and dimension d are written as subscripts, i.e. v_{id} is the dth component of the velocity vector of particle i. The meaning of the velocity equations is also shown in Fig. 1, where the new position directly follows from the addition of the three vectors, namely the velocity component, the cognitive component and the social component.

Algorithm 2 Variant 1 of position / velocity update rule

Input: position $\alpha_i \equiv \mathbf{r}_i$, velocity $\delta_i \equiv \mathbf{v}_i$ and local best position $\beta_i \equiv \mathbf{l}_i$ for each particle i
global best position $\gamma \equiv \mathbf{g}$ and cognitive parameter c_1, social parameter c_2 and weight w
Output: α, with $\alpha_i \in \mathcal{S}$, the solution space, and δ

for all particles i **do**
  $r_1 \leftarrow$ *get_uniform_random_number*$(0,1)$
  $r_2 \leftarrow$ *get_uniform_random_number*$(0,1)$
  for all dimensions d **do**
    $v_{id} \leftarrow w \cdot v_{id} + c_1 \cdot r_1 \cdot (l_{id} - x_{id}) + c_2 \cdot r_2 \cdot (g_d - x_{id})$
    $x_{id} \leftarrow x_{id} + v_{id}$
return $\alpha \equiv \{\mathbf{r}_i\}, \delta \equiv \{\mathbf{v}_i\}$

Each vector in Equ. (1) is weighted by a scaler, which changes only the length of the vectors and not their directions. Depending on the choices of the parameters w, c_1 and c_2 the influence of each component can be adapted. The new position follows then from the superposition of the three vectors as it is also visualized in Fig. 1.

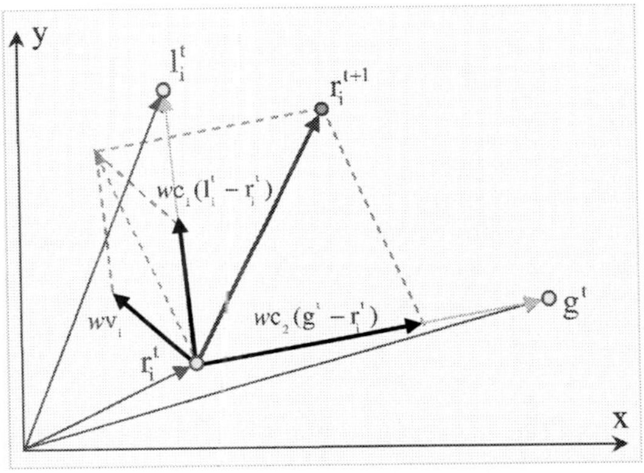

Figure 1. Visualization of the iterative solution update in PSO1 according to Equ. (1).

For this velocity update rule Wilke showed that the search space of the particles is restricted to line search if no new global or local best solution is found. This is a disadvantage because the swarm individuals are not able to explore a sufficiently large area of the problem space [16]. We will see that this behavior leads to a worse performance later on in the experimental Sec. 3., but for the moment we will consider the second variant.

2.2.2. Variant 2

As already mentioned in the previous section the original version of the algorithm is written ambiguously. The second velocity update rule PSO2 is given by

$$\mathbf{v}_i^{t+1} = w \cdot \left[\mathbf{v}_i^t + c_1 \cdot \mathcal{R}_1 \cdot \left(\mathbf{l}_i^t - \mathbf{r}_i^t\right) + c_2 \cdot \mathcal{R}_2 \cdot \left(\mathbf{g}^t - \mathbf{r}_i^t\right)\right], \quad (2)$$

where $\mathcal{R}_1$ and $\mathcal{R}_2$ are two diagonal matrices. The elements are uniform random numbers in $[0,1)$. On the first view this seems quite complicated but looking at the pseudo code it reveals to be only a minor change to the algorithm variant

PSO1. Here the computation of the random numbers is included in the inner loop as can be seen in Algorithm 3.

Algorithm 3 Variant 2 of position / velocity update rule

Input: position $\alpha_i \equiv \mathbf{x}_i$, velocity $\delta_i \equiv \mathbf{v}_i$ and local best position $\beta_i \equiv \mathbf{l}_i$ for each particle i
global best position $\gamma \equiv \mathbf{g}$ and cognitive parameter c_1 and social parameter c_2
Output: α, with $\alpha_i \in \mathcal{S}$, the solution space, and δ

$c \leftarrow c_1 + c_2$
$w \leftarrow 2 / \left|2 - c - \sqrt{c^2 - 4c}\right|$
for all particles i **do**
  for all dimensions d **do**
    $r_1 \leftarrow$ *get_uniform_random_number*$(0,1)$
    $r_2 \leftarrow$ *get_uniform_random_number*$(0,1)$
    $v_{id} \leftarrow w \cdot (v_{id} + c_1 \cdot r_1 \cdot (l_{id} - x_{id}) + c_2 \cdot r_2 \cdot (g_d - x_{id}))$
    $x_{id} \leftarrow x_{id} + v_{id}$
return $\alpha \equiv \{\mathbf{r}_i\}, \delta \equiv \{\mathbf{v}_i\}$

An interpretation of the expression $\mathcal{R}_1(\mathbf{l}_i - \mathbf{r}_i)$ is that each vector component of the vector $(\mathbf{l}_i - \mathbf{r}_i)$ is weighted by a random number. Analogously, the vector $(\mathbf{g} - \mathbf{r}_i)$ is treated. Since every component is multiplied with a different random number the vectors are not only changed in length but also perturbed from their original direction. Hence, this version does no longer show the disadvantage of version 1, namely the collapse of the search to line search [16].

In 2002 Clerc and Kennedy analyzed the PSO algorithm theoretically and proposed a new version within this concept given by the velocity update rule (2) with a weight factor

$$w = \frac{2}{\left|2 - c - \sqrt{c^2 - 4c}\right|} \quad \text{and } c = c_1 + c_2 \,. \tag{3}$$

This algorithm was extensively studied and an optimal parameter setup was determined by Carlisle and Dozier with $c_1 = 2.8$ and $c_2 = 1.3$ [17].

2.2.3. Variant 3

Variant 3 of the velocity update rule is based on the first variant described in Sec. 2.2.1.. The problem of the version PSO1 is that the movements of the particles collapse to line searchers and thus can only cover a small area of the search domain. To overcome this problem the two vectors $(\mathbf{l}_i - \mathbf{r}_i)$ and $(\mathbf{g} -$

$\mathbf{r}_i$) are perturbed by multiplying them with two independent random rotation matrices $\mathcal{Q}_1$ and $\mathcal{Q}_2$ [16]. Thus, the velocity equation is now

$$\mathbf{v}_i^{t+1} = w \cdot \mathbf{v}_i^t + c_1 \cdot r_1 \cdot \mathcal{Q}_1 \cdot \left(\mathbf{l}_i^t - \mathbf{r}_i^t\right) + c_2 \cdot r_2 \cdot \mathcal{Q}_2 \cdot \left(\mathbf{g}^t - \mathbf{r}_i^t\right) . \tag{4}$$

A remaining problem is the construction of such a random rotation matrix $\mathcal{Q}$. A rotation matrix has to be orthogonal, meaning that the row vectors are orthonormal to each other. Thus, also the relation $\mathcal{Q}^T = \mathcal{Q}^{-1}$ is valid and this property is necessary to achieve the invariance of the rotation, because $\mathcal{Q}\mathcal{Q}^{-1} = I$ and therefore a rotation by φ and a succeeding rotation by an angle $-\varphi$ results in the original vector.

A method to construct such a rotation matrix is to use the exponential of a matrix by applying the series method described in [18]

$$\mathcal{Q} = e^{\mathcal{W}} = I + \frac{1}{1!}\mathcal{W} + \frac{1}{2!}\mathcal{W}^2 + \frac{1}{3!}\mathcal{W}^3 + \dots , \tag{5}$$

where $\mathcal{W} = \frac{\varphi\pi}{180°}\left(\mathcal{A} - \mathcal{A}^T\right)$. The expression $\left(\mathcal{A} - \mathcal{A}^T\right)$ represents an antisymmetric matrix, i.e. the matrix elements follow the relation $a_{ij} = -a_{ji}$. The antisymmetric part of a matrix contains only rotational gradients, i.e. changes of the original system due to a rotation. The matrix $\mathcal{Q}$ consists now of a unit matrix I transforming the original system into itself and the asymmetric matrix $\mathcal{W}$ including the rotational changes of the original system.

To obtain a random rotation we compute the random matrix $\mathcal{A}$ with a_{ij} being a random number in $[-0.5, 0.5]$. For our purposes it is sufficient to use only the first two terms of the Taylor series expansion of Equ. (5) because we only have small perturbations, i.e.

$$\mathcal{Q} = I + \mathcal{W} = I + \frac{\varphi\pi}{180°}\left(\mathcal{A} - \mathcal{A}^T\right) . \tag{6}$$

The resulting pseudo code is shown in Algorithm 4. From this we can see that the computation of the rotational matrices is very expensive because we have to compute $2D^2N$ additional random numbers compared to variant PSO1, where d is the problems dimension and N is the number of particles. So when looking at the performance later on we have to keep this fact in mind.

2.2.4. Variant 4

One disadvantage of the previously introduced variants is their fast convergence behavior, which often prevents the swarm individuals from escaping from local

Algorithm 4 Variant 3 of position / velocity update rule

Input: position $\alpha_i \equiv \mathbf{r}_i$, velocity $\delta_i \equiv \mathbf{v}_i$ and local best position $\beta_i \equiv \mathbf{l}_i$ for each particle i
global best position $\gamma \equiv \mathbf{g}$, cognitive parameter c_1, social parameter c_2, weight w, angle φ
Output: α, with $\alpha_i \in \mathcal{S}$, the solution space, and δ

for all particles i **do**
    $r_1 \leftarrow$ *get_uniform_random_number*$(0,1)$
    $r_2 \leftarrow$ *get_uniform_random_number*$(0,1)$
    for all dimension l **do**
        for all dimension k **do**
            $a^1_{lk} \leftarrow$ *get_uniform_random_number*$(-0.5, 0.5)$
            $a^2_{lk} \leftarrow$ *get_uniform_random_number*$(-0.5, 0.5)$
    for all dimension l **do**
        for all dimension k **do**
            $q^1_{lk} \leftarrow 0,\ q^2_{lk} \leftarrow 0$
            if $l = k$ **then**
                $q^1_{lk} \leftarrow 1,\ q^2_{lk} \leftarrow 1$
            $q^1_{lk} \leftarrow q^1_{lk} + \frac{\varphi\pi}{180°}\left(a^1_{lk} - a^1_{kl}\right)$
            $q^2_{lk} \leftarrow q^2_{lk} + \frac{\varphi\pi}{180°}\left(a^2_{lk} - a^2_{kl}\right)$
    for all dimensions d **do**
        $v_{id} \leftarrow w \cdot v_{id}$
        for all dimensions l **do**
            $v_{id} \leftarrow v_{id} + c_1 \cdot r_1 \cdot q^1_{dl} \cdot (l_{id} - x_{id}) + c_2 \cdot r_2 \cdot q^2_{dl} \cdot (g_d - x_{id})$
        $x_{id} \leftarrow x_{id} + v_{id}$
return $\alpha \equiv \{\mathbf{r}_i\}, \delta \equiv \{\mathbf{v}_i\}$

minima in the optimization process. In particular, a known result is that the particles quickly converge to a small area of the search space with a rate of shrinkage following a scaling law for spherical symmetric functions [19].

An idea to overcome this premature convergence is the introduction of a repulsive force among the individuals. The approach of *Repulsive PSO* uses an attraction to the global minimum and a repulsive force depending linearly on the distance of the current particle and a randomly chosen other particle [20]. However, a main disadvantage is that the repulsive force increases with distance, which seems to be unnatural, and the performance improvement was found to be minor. Another version introduces a repulsive force according to the Coulomb law as suggested by Blackwell et al. and is hence named *Charged PSO* (CPSO) [21]. Besides strongly positive effects as good results for dynamic search spaces one faces now the problem to encompass final convergence if applied to static problems. We proposed a solutions to this problem. The resulting algorithm uses a dynamic *charge reduction* over time defining *particle groups* which are iteratively merged, reducing the overall charge of the system gradually during the optimization run and shows superior results in complex state spaces or high dimensionality [22].

The CPSO approach is based on the PSO2 version, where an acceleration term is added to the equations of motion, which are given by

$$\mathbf{v}_i^{t+1} = w\left[\mathbf{v}_i^t + c_1 \cdot \mathcal{R}_1 \cdot \left(\mathbf{l}_i^t - \mathbf{r}_i^t\right) + c_2 \cdot \mathcal{R}_2 \cdot \left(\mathbf{g}^t - \mathbf{r}_i^t\right) + \mathbf{a}_i\right] . \tag{7}$$

The acceleration of a particle i is then determined additively from the repulsive force term between particle i and k by

$$\mathbf{a}_i = \sum_k \mathbf{a}_i(k) . \tag{8}$$

While in the classical model $\mathbf{a}_i = \mathbf{0}$, in CPSO the repulsive term is based on the well-known Coulomb law [21], which is also similar to the commonly used gravitational force in swarm robotics [23]. The easiest choice of the force is

$$\mathbf{a}_i(k) = \frac{Q_i Q_k}{|\mathbf{r}_i - \mathbf{r}_k|^3} (\mathbf{r}_i - \mathbf{r}_k) . \tag{9}$$

In this application the charge product $Q_i Q_k$ is defined to be -1 for particles of different groups and zero otherwise.

The idea now is to modify the number of repelling particles with time by reorganizing particles in groups. On algorithm start, the group number G_0 equals the number of particles P and the behavior is just like in CPSO. If we sum up all pairs which repel each other to a variable S (*charge sum*), we have

$$S_0 = \binom{G_0}{2} = \binom{P}{2} = \frac{P^2 - P}{2} . \tag{10}$$

The subscripts of S_i count the number of reorganizations, and with each reorganization the number of groups is reduced gradually. A straightforward approach is for example to unify pairs of groups in each iteration. Finally after n iterations the dynamic reduction of the number of groups eventually conducts convergence to classical PSO2 and all particles end up in one single group, i.e. $S_n = 0$ and $G_n = 1$. This guaranties an excellent convergence of the algorithm.

A large variety of merging schemes are possible. A highly efficient one is based on integer division and was named *Random CPSO* because the particles can change the group belonging over time instead of new groups being constructed purely by collapsing two or more groups (*Deterministic CPSO*), see

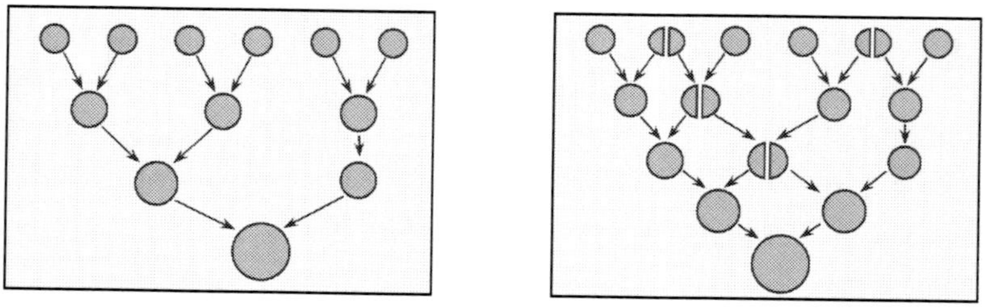

(a) Deterministic CPSO (b) Random CPSO

Figure 2. Visualization of possible grouping and reorganization schemes for CPSO.

Table 1. Random grouping for a swarm with 10 particles *using integer division*

Number of Merges	Group belongings of the particles										Charge sum S
0	0	1	2	3	4	5	6	7	8	9	45
1	0	0	1	1	2	2	3	3	4	4	40
2	0	0	0	1	1	1	2	2	2	3	36
3	0	0	0	0	1	1	1	1	2	2	32
4	0	0	0	0	0	1	1	1	1	1	25
5	0	0	0	0	0	0	1	1	1	1	24
6	0	0	0	0	0	0	0	1	1	1	21
7	0	0	0	0	0	0	0	0	1	1	16
8	0	0	0	0	0	0	0	0	0	1	9
9	0	0	0	0	0	0	0	0	0	0	0

Fig. 2. An example of the group assignment for Random CPSO is shown in Tab. 1 and the pseudo code is shown in Algorithm 5.

The number of merging iterations in *random grouping* is $P-1$, caused intrinsically by the behavior of the integer division. In this case the number of groups in each iteration is not reduced exactly by one (cp. Tab. 1). In the first iterations the number of groups decreases faster but afterwards the group number keeps constant over many iterations, which results in a *quasi constant merging speed*, i.e. an almost constant decrease of the charge sum of the swarm. For the performance comparisons in Sec. 4. the merging speed is chosen in a way that after 50% (i.e. $p = 0.5$) of the allowed number of objective function evaluations all particles are collapsed into one single group.

Having introduced four different realizations of the functions *compute_new_*

Algorithm 5 Variant 4 of position / velocity update rule

Input: position $\alpha_i \equiv \mathbf{r}_i$, velocity $\delta_i \equiv \mathbf{v}_i$ and local best position $\beta_i \equiv \mathbf{l}_i$ for each particle i
global best position $\gamma \equiv \mathbf{g}$, cognitive parameter c_1, social parameter c_2, weight w, and group parameter t

Output: α, with $\alpha_i \in \mathcal{S}$, the solution space, and δ, t

$c \leftarrow c_1 + c_2$
$w \leftarrow 2/|2 - c - \sqrt{c^2 - 4c}|$
for all particles i **do**
  $\mathbf{a}_i \leftarrow \mathbf{0}$
  for $k = i + t$ **to** $k < i + P$ **with** $k = k + t$ **do** {equivalent to i div $t \neq k$ div t}
    $\mathbf{a}_i \leftarrow \mathbf{a}_i + \mathbf{a}_i(k \bmod P)$ {see Equ. (9)}
  for all dimensions d **do**
    $r_1 \leftarrow$ *get_uniform_random_number*$(0, 1)$
    $r_2 \leftarrow$ *get_uniform_random_number*$(0, 1)$
    $v_{id} \leftarrow w \cdot (v_{id} + c_1 \cdot r_1 \cdot (l_{id} - x_{id}) + c_2 \cdot r_2 \cdot (g_d - x_{id}) + a_{id})$
    $x_{id} \leftarrow x_{id} + v_{id}$
if $t < N \wedge$ *fixed_iterations_reached*() **then**
  $t \leftarrow t + 1$
return $\alpha \equiv \{\mathbf{r}_i\}, \delta \equiv \{\mathbf{v}_i\}, t$

state$(\alpha, \beta, \gamma, \delta, w, c_1, c_2)$ we will shortly consider different heuristics and boundary conditions in the next sections.

2.3. Heuristics

In literature one can find the application of several heuristics to PSO in order to further improve the performance [16]. In the following we will introduce several commonly used heuristics and later on we will test some of them whether they are beneficial to the algorithms.

Random Initial Velocities: Each particle will get a random initial velocity, whereas the velocity components are chosen to be within 10% of the search space size. This option increases the diversity of the particles because now they not only differ in their initial position but also in their initial velocities. On the other hand, a random initialization does not consider any characteristics of the search space and the initial velocities are forgotten during an optimization run due to the damping by the weight factor w. Thus, it is questionable whether it is an improvement of the algorithm.

Maximum Velocity Restriction: This is a commonly used heuristic [24] and is introduced to overcome instabilities of the algorithm in case of high velocities. Basically there are two options: either we can restrict each velocity component, i.e. $|v_{id}| \leq v_{\max}$, or we restrict the total velocity vector of each particle i to a given length, i.e. $|\mathbf{v}_i| = \sqrt{\mathbf{v}_i \mathbf{v}_i} < v_{\max}$.

The first option, namely the restriction of the single velocity components, does not only change the length of the velocity vector but also its direction and thus introduces an additional diversity into the algorithm. Another advantage is its simple implementation. The second option only changes the length of the velocity vector and therefore leads to a slower convergence but also to the minimization of instabilities. The impact of both options on a vector is shown in Fig. 3.

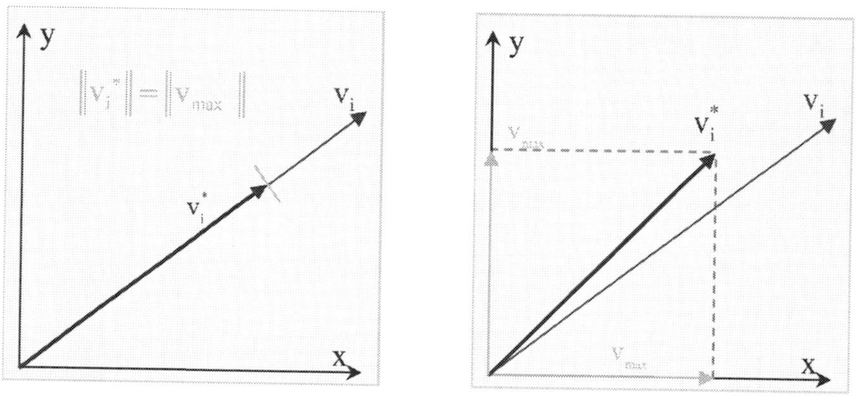

(a) Restriction of vector length (b) Restriction of component size

Figure 3. Visualization of the effect of different maximum velocity restrictions on the length and direction of the velocity vector.

Wilke showed that the application of the first velocity restriction option improves variant PSO1 mainly due to the additional diversity which prevents the particles from collapsing to line searchers [16]. For PSO2 Carlisle and Dozier pointed out that a fine tuning of $v_{\max}$ can help to improve the performance but rather suggested to abandon this option and to introduce adequate boundary conditions, i.e. the velocity of the particle is set to zero at the boundary [17]. More details on the influence of boundary conditions are given in Sec. 2.4..

Minimum Velocity Restriction: We can also restrict the particles' velocities by introducing a minimal velocity in order to prevent the algorithm from early convergence. However, it is questionable whether the algorithm can converge in general. Because upon convergence the swarm contracts to a small region in search space and is able the perform a fine exploration of this region. In this case also the velocities of the individuals become smaller and smaller. But with the application of the minimum velocity restriction, it will not be possible to obtain this behavior.

A possible implementation of these velocity restriction rules, namely the minimum and maximum velocity restriction applied to single vector components, is shown in Algorithm 6.

Algorithm 6 Realization of velocity restriction rules

Input: velocity $\mathbf{v}_i$ for each particle i, minimum velocity $v_{\min}$ and maximum velocity $v_{\max}$
Output: new velocity value $\mathbf{v}_i$ for each particle

1: **for all** particles i **do**
2:     **for all** dimensions d **do**
3:         **if** $|v_{id}| < v_{\min}$ **then**
4:             $v_{id} \leftarrow \text{sgn}(v_{id}) \cdot v_{\min}$
5:         **if** $|v_{id}| > v_{\max}$ **then**
6:             $v_{id} \leftarrow \text{sgn}(v_{id}) \cdot v_{\max}$
7: **return** $\delta \equiv \{\mathbf{v}_i\}$

Craziness: This heuristic places particles with a certain probability p randomly in the search space in the course of the algorithm. A probability $p = 1$ would reduce the particles to random walkers and for $p = 0$ we will get the original behavior of PSO. This heuristic can be compared to the mutation operator in *Genetic Algorithms* because both introduce additional diversity in the algorithm and can help to overcome local minima.

Unfortunately, we have to point out that we will encounter the same problems like for the minimum velocity restriction, namely the failing of the final contraction of the swarm at the best found solution. Similar results are also reported by Wilke, who shows that this heuristic improves variant PSO1 due to the introduction of more diversity but fails to improve PSO2 [16].

Adaptive Weight Factor: Zheng *et al.* suggested to adapt the weight factor w during an optimization run starting at a weight $w = 0.4$ and iteratively increasing it to $w = 0.9$ [25]. They argued that small weights increase the search abilities of the algorithm and large weight values help to stabilize the algorithm. Therefore, this setup allows a good exploration of the search domain in the beginning with $w = 0.4$ and by slowly increasing this value the algorithm converges.

This method is applicable for PSO variants 1 and 3. In the other two variants the weight factor is directly given by the cognitive and social parameters and thus makes the update of w difficult.

Neighborhood: The neighborhood defines which individuals can share information about the global best solution among each other. Up to now we have assumed a global neighborhood, i.e. the information about the global optimum is distributed instantaneously among the complete swarm. Arbitrary other neighborhoods are possible. For instance the restriction to nearest neighbors, where the information can only be passed to the nearest neighbors and thus needs time to be distributed in the whole swarm. This leads to the emergence of groups within the swarm which can search different regions of the search space [26]. The complexity of the neighborhood can by very different, starting from the easy choices of global and nearest neighborhoods up to complex network relations like preferential attachment or Erdős-Renyi networks [26, 27].

Our test suite includes the first three heuristics and the weight factor can be easily controlled by the user [28]. Regarding the neighborhood we would like to point out that Carlisle and Dozier reported the best performance for a global neighborhood [17]. For this reason, the huge amount of possible other neighborhood relations and also the simplicity of implementing a global neighbor relation, we restrict our consideration to the global neighborhood in this chapter.

2.4. Boundary Conditions

Also the boundary conditions can have an influence on the algorithm performance. Therefore, we will describe three different versions of possible boundary conditions, introduced by Robinson and Rahmat-Samii [29]. For a better understanding of their nature we visualize the variants in Fig. 4 and shortly explain them in the following.

Invisible Boundaries: In the case of invisible boundary conditions we do not change anything on the particles' position or velocity at the boundaries. Consequently, the particle is allowed to enter forbidden regions, but we only will evaluate the objective function in the feasible search space. This kind of boundary condition is possible because the PSO algorithm will automatically pull the particle back into the feasible search space due to the attraction of an individual to the global optimum and its local optimum, which obviously are within the search domain.

Absorbing Boundaries: When the particle reaches the search domain boundaries, absorbing boundary conditions lead to the absorbtion of all velocity components which would bring the individual out of the search domain. We can formulate this mathematically by

$$x_{id} = \begin{cases} x_{\min} & x_{id} + v_{id} < x_{\min} \\ x_{id} + v_{id} & x_{\min} \leq x_{id} + v_{id} \leq x_{\max} \\ x_{\max} & x_{id} + v_{id} > x_{\max} \end{cases} , \tag{11}$$

with predefined boundary values $x_{\min}$ and $x_{\max}$ for each component. The velocities are adapted by

$$v_{id} = \begin{cases} v_{id} & x_{\min} \leq x_{id} + v_{id} \leq x_{\max} \\ 0 & \text{otherwise} \end{cases} , \tag{12}$$

where the equation for v_{id} depends on the used PSO variant.

Reflecting Boundaries: With reflecting boundary conditions the particle is reflected at the search domain boundaries by changing the sign of the responsible velocity components and not allowing the particle to enter the forbidden region by restricting its position to the minimum respectively maximum value. The corresponding equation for the position is equivalent to (11) and the velocity equation is

$$v_{id} = \begin{cases} v_{id} & x_{\min} \leq x_{id} + v_{id} \leq x_{\max} \\ -v_{id} & \text{otherwise} \end{cases} , \tag{13}$$

where again v_{id} depends on the used PSO variant.

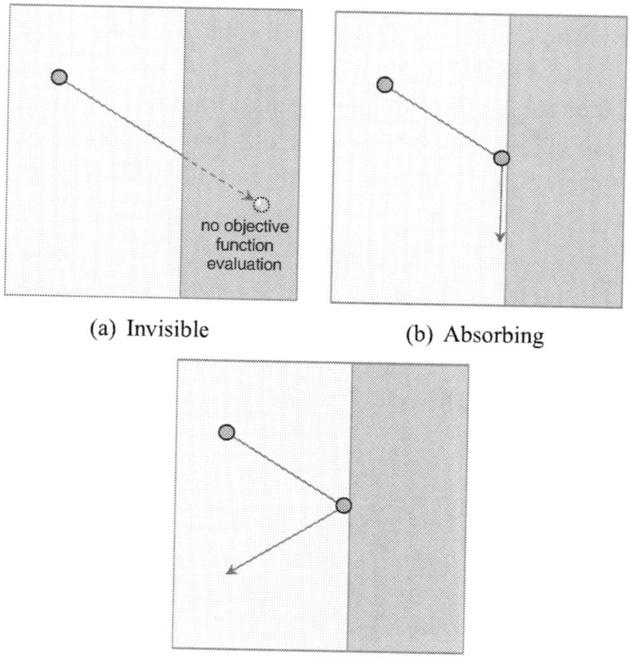

(a) Invisible (b) Absorbing

(c) Reflecting

Figure 4. Visualization of different boundary conditions for a particle movement in a two-dimensional search space.

3. Performance Comparison of Particle Swarm Optimization Approaches

In this section we compare the different PSO variants regarding their performance for continuous benchmark functions in order to determine the method with the best performance for this class of problems.

In order to compare different algorithms we have to introduce *algorithm quality measures*. All in all the solution quality of an optimization heuristic can be measured by a number of different measures [30, 31]. Most common are

Q1) The mean final objective value should be as small as possible.

Q2) The final probability of finding the ground state should be as large as possible.

Q3) The expected number of visits to the ground state should be as large as possible.

Q4) The probability of visiting the ground state during the optimization run should be as large as possible.

Q5) The mean final best-so-far objective value should be as small as possible.

From the theoretical point of view it would be necessary to show that an algorithm performs better than another algorithm for each of these measures. However, in this work we predominantly focus on empirical studies. In this case choice Q2 is problematic because for large problem instances we almost never obtain the optimal solution. This would make it difficult to compare the algorithm, since all of them will have a probability of finding the ground state near or equal to zero. Therefore, we almost do not obtain any statistic information from these measures for the comparison of the performance. Similar considerations rule out the quality measures Q3 and Q4. Except from the fact that these three measures do not deliver enough statistical information, they are also hard to determine in experimental studies.

For these reasons we will concentrate on the measures Q1 and Q5, where the last option is the most common choice. We also determine how often the global optimum is obtained in a given number of runs. This is a rough measure for Q2 and Q3.

3.1. Continuous Benchmark Functions

In literature one can find a manifold of continuous benchmark functions for the determination of the performance and comparison of different optimization methods in continuous solution space. In general one can specify the problem by the following definition.

Definition 3.1 (Continuous Function Problem) (cp. [32])
Having a function $f(\boldsymbol{x})$, which has several local and one or more global optima $f(\boldsymbol{x}^\star)$, and a search domain $\boldsymbol{x} \in S$, we want to find the global optimum $f(\boldsymbol{x}^\star) \leq f(\boldsymbol{x}), \forall \boldsymbol{x} \in S$.

Having a closer look at the various test functions from [32, 33], different versions and domain specifications can be found. Therefore, we give a short overview of the function definitions used here in Tab. 2 including their global minima and domains. Further, some of the functions are plotted in Fig. 5.

Table 2. Selection of d-dimensional functions [32, 33]

Name	Function Definition	Domain and Minimum
Ackley Function	$f(\mathbf{x}) = -20\exp\left[-\frac{1}{5}\sqrt{\frac{1}{N}\sum_{i=1}^{N} x_i^2}\right] - \exp\left[\frac{1}{N}\sum_{i=1}^{N}\cos(2\pi x_i)\right] + 20 + \mathrm{e}$	$x_i \in [-32.768, 32.768]$ Min. Pos.: $x_i^\star = 0$ Min. Value: $f(\mathbf{x}^\star) = 0$
Dixon & Price Function	$f(\mathbf{x}) = (x_1 - 1)^2 + \sum_{i=2}^{d} i(2x_i^2 - x_{i-1})^2$	$x_i \in [-10, 10]$ Min. Pos.: $x_i^\star = 2^{(2^i-2)/2^i}$ Min. Value: $f(\mathbf{x}^\star) = 0$
Function w. Normal Dist. Density	$f(\mathbf{x}) = \sum_{i=1}^{d}\left(ix_i^4 + \frac{20}{\sqrt{2\pi}}\mathrm{e}^{-0.5x_i^2}\right)$	$x_i \in [-1.28, 1.28]$ Min. Pos.: $x_i^{\star 2} = \frac{5}{i\sqrt{2\pi}}\mathrm{e}^{-0.5x_i^{\star 2}}$ Min. Value: $f(\mathbf{x}^\star) = \sum_{i=1}^{d} ix_i^{\star 2}(x_i^{\star 2} + 4)$
Griewangk Function	$f(\mathbf{x}) = 1 + \frac{1}{4000}\sum_{i=1}^{N} x_i^2 - \prod_{i=1}^{N}\cos\left(\frac{x_i}{\sqrt{i}}\right)$	$x_i \in [-600, 600]$ Min. Pos.: $x_i^\star = 0$ Min. Value: $f(\mathbf{x}^\star) = 0$
Levy Function	$f(\mathbf{x}) = \sum_{i=1}^{d-1}(y_i - 1)^2\left[1 + 10\sin^2(\pi y_i + 1)\right] + \sin^2(\pi y_1) + (y_d - 1)^2\left[1 + 10\sin^2(2\pi y_d)\right]$ with $y_i = 1 + \frac{1}{4}(x_i - 1)$	$x_i \in [-10, 10]$ Min. Pos.: $x_i^\star = 0$ Min. Value: $f(\mathbf{x}^\star) = 0$
Michalewicz Function	$f(\mathbf{x}) = -\sum_{i=1}^{d}\sin(x_i)\cdot\sin^{20}\left(\frac{ix_i^2}{\pi}\right)$	$x_i \in [0, \pi]$ Min. Pos.: $x_i^\star = (2.21, 1.57, 1.29, 1.92, ..)$ Min. Value: $f(\mathbf{x}^\star) = -9.6602$ $(d = 10)$
Rastrigin Function	$f(\mathbf{x}) = 10N + \sum_{i=1}^{N}\left(x_i^2 - 10\cos(2\pi x_i)\right)$	$x_i \in [-5.12, 5.12]$ Min. Pos.: $x_i^\star = 0$ Min. Value: $f(\mathbf{x}^\star) = 0$
Rosenbrock Function	$f(\mathbf{x}) = \sum_{i=1}^{N-1}\left[100\left(x_i^2 - x_{i+1}\right)^2 + (1 - x_i)^2\right]$	$x_i \in [-2.048, 2.048]$ Min. Pos.: $x_i^\star = 1$ Min. Value: $f(\mathbf{x}^\star) = 0$
Schaffer Function	$f(\mathbf{x}) = -\sin^2\left(\sqrt{y}\right)/\left(1 + 10^{-3}y^2\right)$ with $y = \sum_{i=1}^{d} x_i^2$	$x_i \in [-100, 100]$ Min. Pos.: $\sum_{i=1}^{d} x_i^{\star 2} = 2.4436009$ Min. Value: $f(\mathbf{x}^\star) = -0.9940049$
Schwefel Function	$f(\mathbf{x}) = -\sum_{i=1}^{d} x_i\cdot\sin\left(\sqrt{\lvert x_i\rvert}\right)$	$x_i \in [-500, 500]$ Min. Pos.: $x_i^\star = 420.9687463$ Min. Value: $f(\mathbf{x}^\star) = -418.9828872\cdot d$
Schwefel's Double Sum Function	$f(\mathbf{x}) = \sum_{i=1}^{d}\left(\sum_{j=1}^{i} x_j\right)^2$	$x_i \in [-65536, 65536]$ Min. Pos.: $\sum_{i=1}^{d} x_i^{\star 2} = 0$ Min. Value: $f(\mathbf{x}^\star) = 0$
Sphere Function	$f(\mathbf{x}) = \sum_{i=1}^{N} x_i^2$	$x_i \in [-5.12, 5.12]$ Min. Pos.: $x_i^\star = 0$ Min. Value: $f(\mathbf{x}^\star) = 0$
Step Function	$f(\mathbf{x}) = \sum_{i=1}^{N}\lfloor x_i\rfloor + 6N$	$x_i \in [-5.12, 5.12]$ Min. Pos.: $x_i^\star \in [-5.12, -5.0]$ Min. Value: $f(\mathbf{x}^\star) = 0$
Trid Function	$f(\mathbf{x}) = \sum_{i=1}^{d}(x_i - 1)^2 - \sum_{i=1}^{d-1} x_i\cdot x_{i+1}$	$x_i \in [-100, 100]$ Min. Pos.: $x_i^\star = i(d + 1 - i)$ Min. Value: $f(\mathbf{x}^\star) = -d(d+4)(d-1)/6$

3.2. Parameter Setup

To apply the different variants of the PSO algorithm to a selection of test functions introduced in the previous section, we first give an overview over the parameter setup used for the experiments. The values are oriented at good setups proposed in literature and for all four variants summarized in Tab. 3.

If not given different elsewhere, we assume a population size of 30 and a problem dimension of 10. Additionally, we used for all variants randomly

[1]The value of the weight factor w follows directly from Equ. (3).

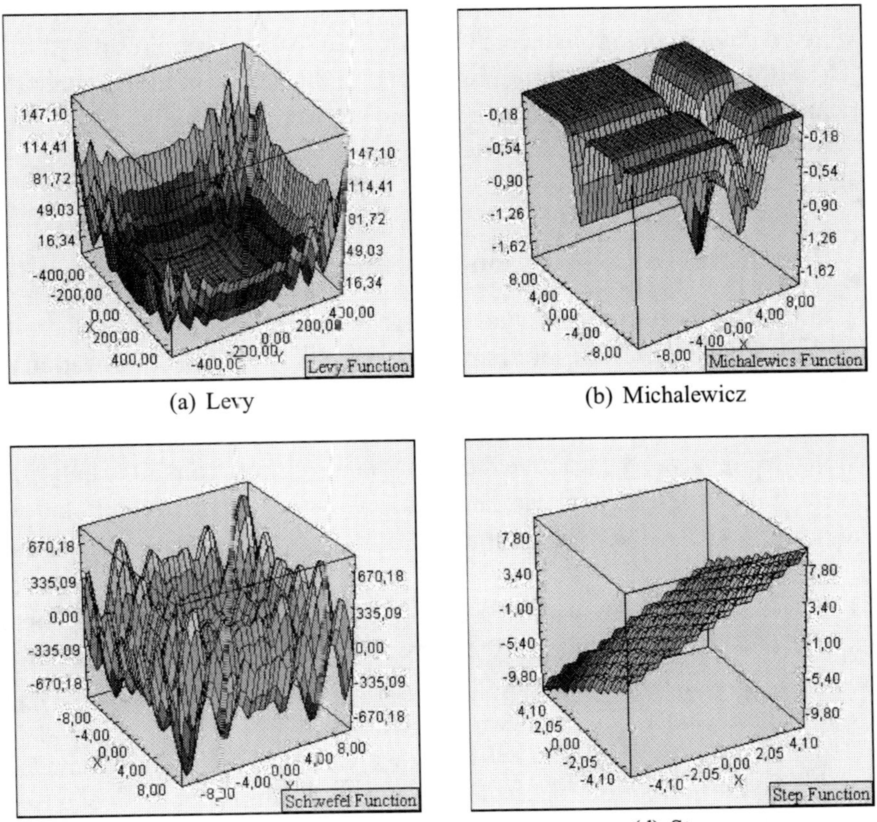

(a) Levy (b) Michalewicz

(c) Schwefel (d) Step

Figure 5. Three-dimensional plots of a selection of continuous benchmark functions.

Table 3. Parameter setup for PSO variants

Parameter	PSO1	PSO2	PSO3	PSO4
weight w	0.8	0.73[1]	0.6	0.73[1]
cognitive parameter c_1	2.0	2.8	2.0	2.8
social parameter c_2	2.0	1.3	2.0	1.3
additional parameters	–	–	$\varphi = 3.0$	$p = 0.5$
references	[12]	[17]	[16]	[22]

distributed initial positions $\mathbf{r}_i$ in the solution space, zero initial velocities $\mathbf{v}_i$ and *invisible boundary conditions*, which are reported to show good results [29].

The different PSO algorithms are applied to the selection of test functions, summarized in Tab. 2. For PSO2 the parameter setup is oriented at [17] with $c_1 = 2.8$, $c_2 = 1.3$, which is reported to outperform the standard setup with $c_1 = c_2 = 2.05$.

3.3. Performance Comparison for Continuous Test Functions

All results for the selected test functions are summarized in Tab. 4, where the best solution and the average solution for 20 runs are given. An optimization run terminates when the global optimum is approximated by 10^{-6} or at most 20,000 function evaluations are done. The results are summarized for the different functions in Tab. 4, where we give the best found solution from all twenty runs, the average best solution of these runs and how often the global optimum was found. Additionally, the average number of objective function evaluations is given for all runs.

The results in Tab. 4 show that PSO converges very quickly to the global optimum for several test functions, especially the Ackley, Levy and Sphere function. For other functions we do not obtain the global optimum and the results also show significant differences between the algorithms.

Comparing the different PSO variants we clearly see the shortcomings of variant PSO1 to find an optimal solution, which is caused by the collapse of the movement to line search as described in Sec. 2.2.1.. The variants PSO2 and PSO4 instead show a good behavior and seem to be quite equivalent in their performance, which is not unusual since both represent the same algorithm in the limit when all groups are merged in PSO4.

Table 4. Performance comparison of different PSO variants in 10 dimensions

	Function	Average Solution	Best Solution	# Successful Runs	# of Function Evaluations
PSO1	Ackley	5.517798	2.583130	0	20000
	Dixon & Price	1.874654	0.262122	0	20000
	FWNDD	71.494089	71.475196	0	20000
	Griewangk	1.052350	0.127008	0	20000
	Levy Function	2.364631	0.257143	0	20000
	Michalewicz	-7.032929	-8.225467	0	20000
	Rastrigin	17.225120	10.810529	0	20000
	Rosenbrock	7.931421	3.711308	0	20000
	Schaffer	-0.025961	-0.065927	0	20000
	Schwefel	-2484.4861	-3221.4607	0	20000
	Double Sum	3363016.3	4812.62	0	20000
	Sphere	0.004895	0.000028	0	20000
	Step	-31.900000	-43.000000	0	20000
	Trid	-173.2207	-209.8003	0	20000
PSO2	Ackley	0.000000	0.000000	20	8310
	Dixon & Price	0.450000	0.000000	2	18690
	FWNDD	71.474933	71.474933	20	4260
	Griewangk	0.055521	0.009865	0	20000
	Levy Function	0.000000	0.000000	20	4440
	Michalewicz	-9.221643	-9.613478	0	20000
	Rastrigin	4.377886	0.995682	0	20000
	Rosenbrock	3.275460	0.795569	0	20000
	Schaffer	-0.791423	-0.994007	7	15390
	Schwefel	-3684.7215	-4071.3905	0	20000
	Double Sum	0.000000	0.000000	19	18090
	Sphere	0.000000	0.000000	20	3720
	Step	-59.550000	-60.000000	12	14730
	Trid	-209.9853	-210.0000	12	18060
PSO3	Ackley	2.492030	0.000000	4	18030
	Dixon & Price	0.450000	0.000000	2	19320
	FWNDD	71.474933	71.474933	20	4470
	Griewangk	0.282767	0.054157	0	20000
	Levy Function	1.834376	0.089528	0	20000
	Michalewicz	-7.800517	-8.990948	0	20000
	Rastrigin	15.824239	6.764713	0	20000
	Rosenbrock	0.405592	0.003044	0	20000
	Schaffer	-0.612497	-0.682340	0	20000
	Schwefel	-2600.3626	-3163.3250	0	20000
	Double Sum	0.000000	0.000000	20	12660
	Sphere	0.000000	0.000000	20	4530
	Step	-45.700000	-53.000000	0	20000
	Trid	-210.000000	-210.000000	20	8340

The variant PSO3 is also able to find good solutions to some of the continuous test functions but shows big differences to the algorithm variants PSO2 and PSO4. For instance PSO3 has trouble to optimize the Ackley and the Levy

Table 4. (Continued)

	Function	Best Solution	Average Solution	# Successful Runs	# of Function Evaluations
PSO4	Ackley	0.000022	0.000000	16	15090
	Dixon & Price	0.475000	0.000000	1	19860
	FWNDD	71.474933	71.474933	20	13380
	Griewangk	0.055168	0.022151	0	20000
	Levy Function	0.000000	0.000000	20	12090
	Michalewicz	-9.222875	-9.581427	0	20000
	Rastrigin	4.037273	0.994959	0	20000
	Rosenbrock	4.061816	1.532471	0	20000
	Schaffer	-0.853756	-0.994007	11	13170
	Schwefel	-3730.8773	-4071.3905	0	20000
	Double Sum	0.000000	0.000000	19	18450
	Sphere	0.000000	0.000000	19	13290
	Step	-58.800000	-60.000000	9	18600
	Trid	-209.9419	-210.0000	2	19980

function unlike to the other two algorithms and on the other hand shows very good results for the Rosenbrock and the Trid function, where the variants PSO2 and PSO4 do not perform so well. However, during the computation we also noticed the additional computation of random numbers for PSO3 and of the repulsive force for PSO4 leading to an increased computation time compared to the other methods. This will be one important point for the comparison of the algorithms later on.

Having not been able to identify a superior method of the four PSO variants with respect to this test functions, we do some more experiments in the next section.

3.4. Investigation of the Convergence Behavior

In this section we investigate the influence of the dimensionality and the population size on the algorithm performance. To realize this we will restrict our considerations to only three test functions which showed distinctive behaviors for the four algorithms in the test runs which are reported in Tab. 4.

- The Levy function is chosen since the algorithms PSO2 and PSO4 were easily able to find the global optimum with about the same number of objective function evaluations, whereas for the algorithm PSO3 we obtained the global optimum in only one run.

- The second function for our tests is the Rastrigin function. None of the four algorithms was able to find the global optimum of this function in ten dimensions but we can use the best and the average objective function values of the twenty runs for a comparison of the algorithms, where these values differed noticeable for the different variants.
- The Rosenbrock function is chosen because the algorithm PSO3 performs here significantly better compared to the other algorithms.

Population size: A widely asked question deals with the population size. Clearly, a larger number of individuals allows to explore the solution space in more detail but on the cost of a higher computational effort. Thus, we have to carefully adjust the population size to get a good performance. Carlisle and Dozier reported the best results for 30 individuals for the algorithm PSO2 [17] and this is also a commonly used population size for the other algorithm variants [16]. We want to test this result for the above given selection of benchmark functions from Tab. 4, i.e. we restrict our consideration to the Levy, Rastrigin and Rosenrock function. The results regarding the population size are summarized in Tab. 5.

For the variants PSO2 and PSO4 the optimal population size seems to be around 30 to 40 when averaging over all three test functions. This is in good agreement with the results of Carlizle and Dosier [17]. A similar result is also obtained for the variant PSO3 with about 40 particles. Regarding the algorithm performance we can see similar results as before, namely the good performance of the variants PSO2 and PSO4 for the Levy and Rastrigin function and the algorithm PSO3 performed very good for the Rosenbrock function.

The greatest discrepancy regarding the population size is obtained for variant PSO1, where the best results were computed for large population sizes. Especially for the Levy and Rosenbrock function the best result is given for a population size of 50 or 60. This behavior can be explained by the collapse of the movement to line search in the case that no new local or global optimum is found. Therefore, the high number of random walkers can prevent this behavior to a certain degree and also can cover a larger amount of the search space. Nevertheless, the results for PSO1 are not as good as for the other algorithms even for a population size of 60.

Table 5. Average results and number of discovered global optima over 20 runs for the PSO variants depending on the population size

	N	PSO1		PSO2		PSO3		PSO4	
Levy	5	3.401786	0	2.558342	0	5.435948	0	0.006889	6
	10	3.579758	0	0.395127	8	4.113506	0	0.090743	18
	20	2.038136	0	0.031669	17	2.293984	0	0.000102	19
	30	2.364631	0	0.000000	20	1.834876	0	0.000000	20
	40	1.808642	0	0.000000	20	1.486812	2	0.000000	20
	50	1.375513	0	0.000000	20	1.356367	3	0.004743	16
	60	1.498357	0	0.000000	20	1.411355	1	0.000452	16
Rastrigin	5	55.228434	0	9.949586	0	27.229960	0	10.288322	0
	10	37.455299	0	8.656167	0	23.232239	0	6.964771	0
	20	21.449210	0	6.765715	0	21.988554	0	4.881877	0
	30	17.225120	0	4.377886	0	15.824239	0	4.037273	0
	40	16.178818	0	3.384456	0	21.093009	0	4.602819	0
	50	22.317950	0	3.587451	0	17.212759	0	6.661133	0
	60	19.463926	0	3.708781	0	18.854436	0	6.983927	0
Rosenbrock	5	231.303016	0	29.558289	0	7.213911	0	7.404082	0
	10	52.901487	0	6.444312	0	1.007168	0	6.283546	0
	20	9.687132	0	3.789316	0	0.606198	0	4.689887	0
	30	7.931421	0	3.275460	0	0.405592	0	4.061816	0
	40	7.302739	0	3.437803	0	0.618989	0	4.640575	0
	50	7.538420	0	3.660280	0	0.820444	0	5.041940	0
	60	7.416475	0	3.908871	0	0.869289	0	6.024519	0

Dependence on Dimensionality: Another important point is the performance of the algorithms depending on the dimensionality of the problem. In order to allow the algorithms to find adequate optimization results also in higher dimensions, we adjusted the maximum number of allowed objective function evaluations to be $2000 \cdot d$ in d dimensions. The results for the three chosen benchmark functions are summarized in Tab. 6. From this table we can see some interesting behaviors which can help us to identify superior behavior.

We already ruled out the algorithm PSO1 due to its worse performance, which can be again seen from these results. For the previously similar behavior of the algorithms PSO2 and PSO4, we now find that PSO4 performs better with increasing dimensionality. As expected for higher dimensions the algorithm solutions become worse but for the algorithm PSO4 this process is much slower

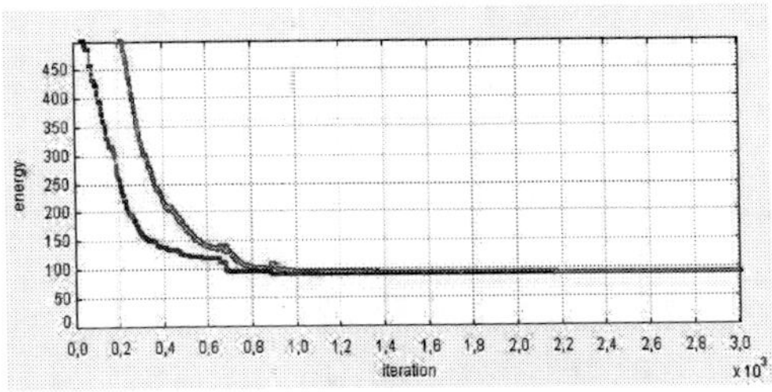

(a) PSO2

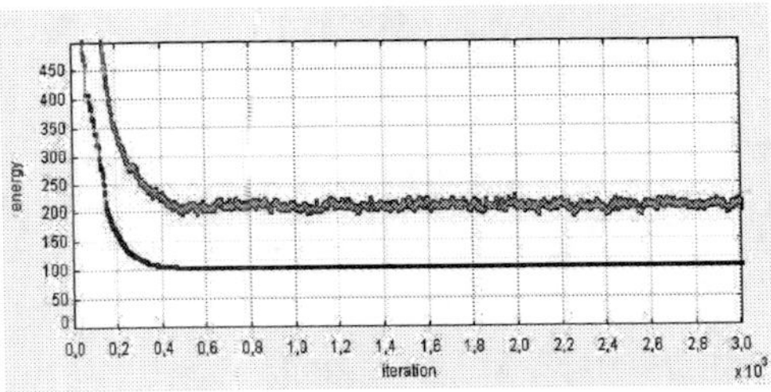

(b) CPSO

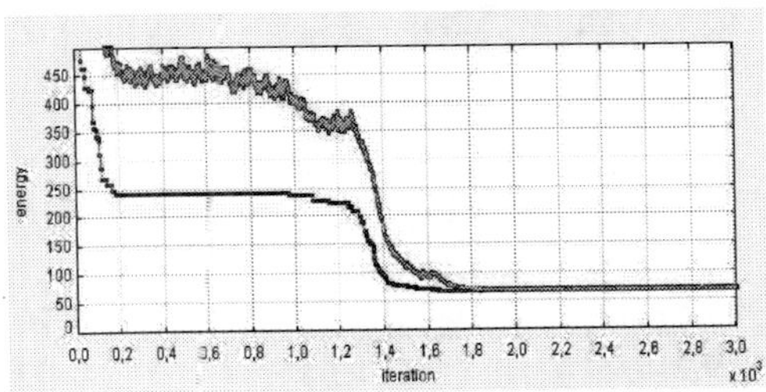

(c) Deterministic CPSO

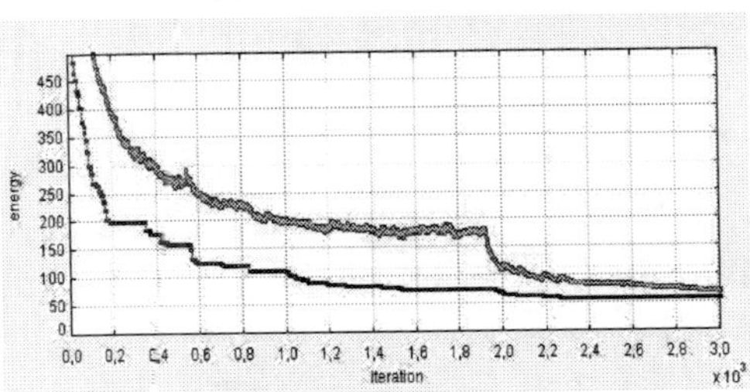

(d) PSO4 (Random CPSO)

Figure 6. Convergence Behavior of different PSO variants: the red line indicates the best so far value and the green line the current average value over the whole swarm (see [22]).

Table 6. Average results over 20 runs for the PSO variants depending on the dimensionality

	D	PSO1	PSO2	PSO3	PSO4
Levy	2	0.000000	0.000000	0.000000	0.000015
	5	0.112268	0.000000	0.256649	0.001136
	10	2.364631	0.000000	1.834876	0.000000
	15	4.075384	0.143859	4.529678	0.000025
	20	6.295161	0.558245	6.771720	0.035542
	25	11.126197	1.386592	9.929187	0.148262
	30	11.601619	2.194235	10.481226	0.663576
	40	19.356817	6.807962	16.191035	2.702797
	50	27.892140	11.452513	22.006409	7.412729
Rastrigin	2	0.310168	0.000000	0.149244	0.179326
	5	5.778121	0.649773	6.050416	1.229930
	10	17.225120	4.377886	15.824239	4.037273
	15	34.287237	9.502961	28.107544	9.854636
	20	58.688028	22.088305	42.832892	16.866711
	25	92.760770	39.883554	58.005958	26.151316
	30	117.669800	48.454424	68.154538	24.648357
	40	193.617488	93.333555	108.052266	62.263781
	50	276.711109	119.127738	125.961506	97.578342
Rosenbrock	2	0.000004	0.000002	0.000000	0.000093
	5	1.178776	0.418176	0.786169	0.539374
	10	7.931421	3.275460	0.405592	4.061816
	15	19.244341	9.290394	1.329256	9.473437
	20	38.064974	14.848237	8.200706	14.368681
	25	67.278711	22.725736	16.414125	19.372121
	30	104.720292	40.952690	23.451070	31.103120
	40	180.182921	70.631538	35.163330	37.103120
	50	257.749798	92.944353	44.817162	57.187222

than for PSO2 and thus the results in 50 dimensions already differ significantly. This behavior is also visualized in Fig. 6, where the best so far solution as well as the current average solutions of the whole swarms are shown for different methods. This includes PSO2 and PSO4, but also CPSO (i.e. without merging of groups) and Deterministic CPSO (i.e. the pairwise collapsing of groups). While PSO2 converges very fast to a local optimum and the average state of CPSO does not converge at all, for the CPSO approaches with merging the convergence is postponed and better results are obtained. In the latter case the results for the almost constant reduction of the charge sum in Random CPSO (PSO4) are better compared to Deterministic CPSO.

Regarding the comparison of the algorithms PSO3 and PSO4 the result is not so obvious because PSO3 yields better results for the Rosenbrock function and worse results for the Levy and Rastrigin function. This gives the algorithm PSO4 a small advantage compared to PSO3 but the main criteria for the preference of PSO4 is its significantly faster computation. Tests showed that for PSO4 already in 30 dimensions the computation was more than five times faster compared to PSO3. The difference is caused by the computation of the rotational matrices, which requires among others $O(d^2N)$ additional random numbers in PSO3, which requires more time as the computation of the repulsive force in PSO4.

As a result PSO2 converges very quickly and finds good solutions to problems of small dimensionality and PSO4 returns better results in complex search spaces with an acceptable increase in running time.

Visualization of the Convergence Process: It can be useful to visualize the movement of the particles in the solution space in real time to get an insight into the convergence behavior.[2] The visualization shows that the swarm slowly concentrates more and more around the global optimum during an optimization run. Thus, the swarm is able to explore the region around the optimum very well and therefore finds in most cases the best possible solution. An exemplary behavior of the individuals is shown in Fig. 7 for the Ackley function in ten dimensions, where all plots show the individuals in the plane (x_0, x_1) of the ten-dimensional space. Red dots are the current positions, blue circles the local best solutions and the green circle is the global best solution.

The visualization can also help us to find reasons for the bad convergence behavior for some of the test functions as shown in Fig. 8 for the Rastrigin and Rosenbrock function. For instance most of the vector components of the final solution of the Rastrigin function have reached their optimal value but there are still one or more vector components in a neighboring local minima. For the Rosenbrock function instead we can see that the swarm converges very slowly and is still distributed along the valley of the Rosenbrock function, where the individuals mostly search at the slopes of the valley and only slowly move to

[2]The update frequency can be specified as simulation parameter to save computation time.

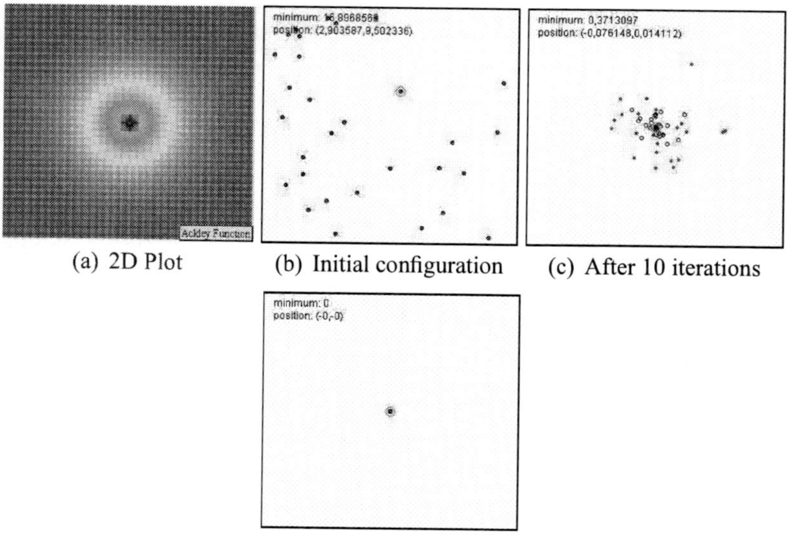

(a) 2D Plot (b) Initial configuration (c) After 10 iterations

(d) Final configuration

Figure 7. Visualization of the search behavior of PSO for the Ackley function.

the global optimum. Thus, with a limited number of function evaluations the swarm has not the change to converge at all.

3.5. Influence of Heuristics and Boundary Conditions

In this section we determine the influence of different heuristics described in Sec. 2.3. on the optimization results. We again restrict our considerations to the Levy, Rastrigin and Rosenbrock function and test the performance of all four algorithm variants for a selection of heuristics introduced in Sec. 2.3. and different boundary conditions from Sec. 2.4.. The results are shown in Tab. 7 and the basic insights are summarized below.

- The application of the *initial velocity restriction* includes the choice of random velocities with a size of 10% of the search space domain. This heuristic has only a minor influence on the performance of the algorithm for the variants PSO2, PSO3 and PSO4, where we observed no improvement or even a decline of the performance in all cases. In the case of the PSO1 algorithm this heuristic is beneficial due to the introduction of an additional diversity into the algorithm in order to prevent the collapse to

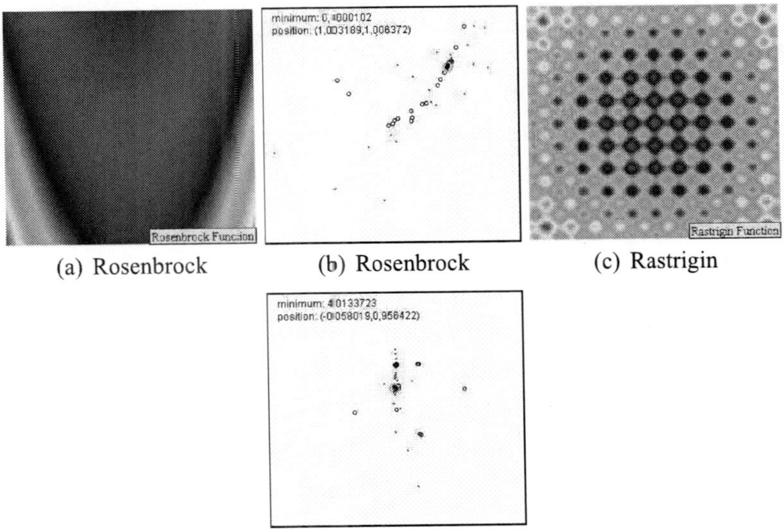

(a) Rosenbrock (b) Rosenbrock (c) Rastrigin

(d) Rastrigin

Figure 8. Visualization of the search behavior of PSO for different functions.

line search. Unfortunately, the influence of this heuristic ranges only to the first iterations since the system will steadily forget its initial state and thus the improvement of PSO1 is minor.

- The *minimum velocity restriction* restricts the velocities to a minimal value which is in our case chosen to be equal to $v_{\min} = 0.01$. This heuristic worsens the performance of the algorithms PSO2, PSO3 and PSO4. Even for PSO1 there was almost no improvement of the performance in all cases since the performance gain due to the introduction of additional diversity to this algorithm variant was not predominant to the disadvantages of this heuristic, namely the prevention of the swarm to finally contract at the so far global optimum.
- The *maximum velocity restriction* limits the velocity components to a $v_{\max}$ value which was chosen to be one eighth of the search space size. This size was suggested by Carlisle and Dosier [17]. However, we did not obtain any improvement of the results for the algorithms PSO2, PSO3 and PSO4. As expected, the algorithm PSO1 was improved significantly since also the maximum velocity prevents the collapse to line search.

- We applied a number of different boundary conditions. For all experiments before we used *invisible boundary conditions* because this is automatically given due to the domain specifications of the test functions and by not evaluating them outside of the search domain. The previously obtained results are again shown in the table for a better overview of the results.

- The *absorbing boundary conditions* sometimes improve the performance and sometimes not. Basically, there is no improvement for the variant PSO2, the results for PSO3 are ambiguous and we get a slight improvement for PSO4 as well as a clear improvement for algorithm PSO1. These boundary conditions yield the best results for PSO1 and thus should be used in connection with its application. Regarding the variant PSO4 the improvements are minor and could also be due to statistical noise in the tests.

- Bad convergence is obtained for *reflecting boundary conditions* for all four algorithms. This is most likely caused by not resetting the velocity of the particle. Therefore, the particle has a very high velocity and randomly wanders often in the outer regions rather then converging to a good solution.

All in all, the greatest influence of boundary conditions has been obtained for the algorithm PSO1, where *absorbing boundary conditions* clearly give the best performance increase. However, it cannot compete with the results from the other variants. The influence of the heuristics and boundary conditions to the other three methods are smaller. For PSO2 and PSO3 no general improvement is found and one should use the plain algorithms instead of spending additional computational efforts for calculating the heuristics. For PSO4 absorbing boundary conditions show slightly better results. However, further testing is required to ensure this behavior.

From our results we propose that the best method for continuous problems presented by this set of test functions of low dimensionality is PSO2 in combination with *invisible boundary conditions* and a parameter setup as given in Sec. 3.2.. For more complex solution spaces it can be beneficial to postpone the fast convergence of PSO2 for example by an repulsive force as in PSO4, which shows the best performance for higher dimensions.

Table 7. Average results over 20 runs for the PSO variants depending on several heuristics and boundary conditions

	Heuristics	PSO1	PSO2	PSO3	PSO4
Levy	initial velocities	1.543018	0.000000	1.890229	0.000000
	minimum velocity	1.417102	0.022792	1.372000	0.022785
	maximum velocity	0.452399	0.000000	2.068814	0.000000
	invisible boundaries	2.364631	0.000000	1.834876	0.000000
	absorbing boundaries	0.100688	0.027194	0.758856	0.000000
	reflecting boundaries	9.835597	0.351737	3.277196	0.048566
Rastrigin	initial velocities	16.921589	4.477325	18.456456	3.581899
	minimum velocity	19.545045	4.172876	19.850581	3.728923
	maximum velocity	13.209399	4.678266	18.236328	3.683502
	invisible boundaries	17.225120	4.377886	15.824239	4.037273
	absorbing boundaries	6.120772	4.576811	17.411761	3.284141
	reflecting boundaries	72.858582	11.323256	30.223059	15.873847
Rosenbrock	initial velocities	7.546446	3.102712	0.607351	3.298437
	minimum velocity	5.319546	5.368353	0.563634	5.448133
	maximum velocity	0.518785	3.074799	0.605660	3.578654
	invisible boundaries	7.931421	3.275460	0.405592	4.061816
	absorbing boundaries	5.280166	3.490276	0.607636	3.552367
	reflecting boundaries	114.336727	4.562828	14.195142	46.223531

4. Comparative Analysis to Alternative Methods

While the focus of our studies in this chapter is to compare different approaches to particle swarm optimization and to identify design characteristics, which seem to be superior compared to others, it is also important to know how good or how bad the PSO approach performs compared to other standard optimization heuristics because this may be essential for the choice and decisions of a practitioner.

Hence, in this section we take different other meta-heuristics into account and compare their performance with the performance of the PSO approach for exactly the same problems as investigated above.

4.1. Global Optimization Heuristics

For the solution of these problems, i.e. benchmark functions in the continuous domain, one can choose between a large number of *general purpose optimization heuristics* [34] (often called *metaheuristics*). An obvious disadvantage of this algorithm class is that the performance can be arbitrary bad, which means we do not know, how far from the optimal solution we are still away. But the advantage of these methods is that they perform very good in practice and can find "good" solutions in moderate time. These methods can be classified in two major categories:

1. There are relatively simple iterative improvement approaches as *Simulated Annealing* (SA) [7], *Threshold Accepting* (TA) [8] or the *Great Deluge Algorithm* (GD) [35] which deal only with one current solution and one candidate solution at the same time but,
2. there are also more involved techniques which deal with more than one candidate solution at the same time as *Genetic Algorithms* (GA) [1] and *Evolution Strategies* [11] or newer paradigms as *Ant Colony Optimization* [3] or, as introduced in the paragraphs before, *Particle Swarm Optimization* (PSO) [4]. Also ensemble-based techniques of the simple iterative approaches as *Ensemble-based Simulated Annealing* or *Ensemble-based Threshold Accepting* (EBTA) [9] are representatives of this second group of algorithms.

A general definition of *global optimization heuristics* is, that these are iterative methods which perform a random walk in a solution space $\mathcal{S}$ based on one state α or a vector of states $\boldsymbol{\alpha}$ to find a global minimum or minimizer. Further, these methods are understood to be of the $\mathcal{A}(0)$ type, utilizing only function values of the objective function but no derivations.

4.2. Single State Methods

Algorithm 7 is a template for a class of methods, which deal only with one current solution and one candidate solution, also called Monte Carlo-type techniques. The degrees of freedom within this template are the *initial state*, the *temperature schedule*, the *move class*, the *acceptance condition* and the *terminating condition*. Comparing different methods, the initial state can be chosen at random identically for all methods and the terminating condition is a fixed number of iterations. For different designs of $P_{\beta\alpha}$, the algorithm shows a different behavior. This is essentially equivalent to the choice of the strategy to get new states from the old ones. Different methods are described by Equ. (14) to (18).

Focusing on this design criterion for algorithms which operate only on one state, the simplest approach besides *random search* with

$$P^{(RA)}_{\beta\alpha} = 1 \tag{14}$$

is to apply the *greedy algorithm* using transition probabilities

$$P^{(GR)}_{\beta\alpha} = \begin{cases} 1 & \mid \text{if } \Delta f \leq 0 \\ 0 & \mid \text{if } \Delta f > 0\,. \end{cases} \tag{15}$$

Algorithm 7 Monte Carlo-type techniques

Input: problem $\mathcal{P}$ with a solution space $\mathcal{S}$

Output: solution $\alpha_{\text{final}} \in \mathcal{S}$

$\alpha \leftarrow$ generate_initial_state($\mathcal{S}$)
for $t \leftarrow 1$ **to** ∞ **do**
    $T \leftarrow$get_new_temperature(α, t, T)
    $\beta \leftarrow$get_new_state($\alpha, t, \mathcal{S}$)
    acceptance_condition $\leftarrow$ *true* with probability $P_{\beta\alpha}$, *false* otherwise
    if (*acceptance_condition*) **then**
        $\alpha \leftarrow \beta$
    if (*terminating_condition*) **then**
        return α

The parameter $\Delta f = f(\beta) - f(\alpha)$ is the difference between the objective function values of the new and the current state. A general disadvantage of the greedy algorithm is that it can get stuck in local optima. It is well known that Monte Carlo-type optimization algorithms like SA or TA are able to overcome local optima. For SA one uses then

$$P^{(SA)}_{\beta\alpha} = \begin{cases} 1 & \mid \text{if } \Delta f \leq 0 \\ e^{-\Delta f/(k_B \cdot T)} & \mid \text{if } \Delta f > 0\,, \end{cases} \tag{16}$$

where a temperature parameter T is introduced, which gets in general reduced during the algorithm execution. For TA one uses the simpler step function for $P_{\beta\alpha}$:

$$P^{(TA)}_{\beta\alpha} = \begin{cases} 1 & \mid \text{if } \Delta f \leq T \\ 0 & \mid \text{if } \Delta f > T\,. \end{cases} \tag{17}$$

Franz et al. [36] were able to prove that Equ. (17) is an optimal decision rule for this kind of algorithm (but this depends on the choice of other parameters, especially on the choice of the temperature schedule, for which the optimal choice is in practice unknown). A further simplification of TA results in the Great Deluge Algorithm with

$$P_{\beta\alpha}^{(GD)} = \begin{cases} 1 & | \text{ if } E(\beta) \leq T \\ 0 & | \text{ otherwise} \end{cases} . \tag{18}$$

4.3. Temperature Parameter

For Monte Carlo-type algorithms there have been several strategies for the temperature schedule investigated in literature. Especially for the ensemble based strategies there are adaptive schedules which show excellent performance, as e.g. *constant thermodynamic speed scheduling* [10]. For comparative studies it is more convenient to use a temperature schedule which realizes a temperature reduction to a target temperature in a fixed number I of available iterations. This is necessary for a meaningful comparison. A standard technique to do this is to use an exponential schedule

$$T_{\text{new}} = c \cdot T \tag{19}$$

in each iteration step, where the parameter c is chosen as

$$c = \exp\left[\ln\left(\frac{T_{\text{final}}}{T_0}\right)/I\right] . \tag{20}$$

In the setup we used an adaptively obtained start temperature using

$$T_0 = c \cdot \sigma_{E|T=\infty} = \sqrt{\langle E^2\rangle_{T=\infty} - \langle E\rangle_{T=\infty}^2} , \tag{21}$$

where c is a small constant, except for GD, where higher values are necessary. The energies at $T = \infty$ are obtained by realizing an unbiased random walk with the acceptance rule as given by Equ. (14). As temperature schedule we used exponential cooling according to Equ. (19) and (20), where I is an in advance fixed number of allowed steps of the method. The final temperature has been chosen to be $T_{\text{final}} = 10^{-7}$ in general but to a different value if this improved the results for the given function.

4.4. Move Classes for the Continuous Domain

Applying different move classes means in the continuous domain basically to apply different probability distributions to choose the step width of the next step. One possibility is the variation of single dimensions of the current solution by a random value. There are several distribution possible and we will present here the three used by us.

Uniform-Random-Move: Changes one coordinate x_i of state $\alpha = \mathbf{x}$ by a uniform random value in the range $-max\_step$ to max_step. This is the simplest move class within this context and only needs numbers from a standard uniform random number generator and a user defined value for *max_step*.

Gauss-Random-Move: Changes one coordinate x_i of state $\alpha = \mathbf{x}$ by a Gaussian random number with standard derivation $\sigma = max\_step$ and expectation value $\mu = x_i$. The Gaussian probability distribution is given by

$$\mathcal{N}(\mu,\sigma) = \frac{1}{\sqrt{2\pi\sigma^2}} \exp\left[-\frac{(x-\mu)^2}{2\sigma^2}\right] . \tag{22}$$

The random numbers are generated from uniform random numbers u using the Box-Muller method [37], which enables the generation of $\mathcal{N}(0,1)$ distributed number and is given by

$$\begin{aligned} r_1 &\leftarrow \sqrt{-2\ln(1-u_1)} \cdot \cos(2\pi u_2) \\ r_2 &\leftarrow \sqrt{-2\ln(1-u_1)} \cdot \sin(2\pi u_2) \end{aligned} .$$

Due to the relation $\mathcal{N}(\mu,\sigma) \leftarrow \mu + \sigma r$, where r is Gaussian distributed with $\mathcal{N}(0,1)$, we can easily compute the desired numbers.

Lorentz-Random-Move: Changes one coordinate x_i of state $\alpha = \mathbf{x}$ by a random number generated from the Cauchy-Lorentz distribution

$$\mathcal{L}(s,t) = \frac{1}{\pi} \frac{s}{s^2 + (x-t)^2} . \tag{24}$$

For the Cauchy-Lorentz distribution all moments are not defined and thus this distribution has no expectation value, standard deviation and variance. However, the parameters s and t still have a meaning. The median of the distribution is given by $x_m = t$ and additionally for x_m the function reaches its maximum value given by $f(x_m) = 1/(\pi s)$.

To generate random numbers from the Cauchy-Lorentz distribution, we can use the inversion method for which we need the cumulative distribution function

$$F(x) = \frac{1}{2} + \frac{1}{\pi} \arctan \frac{x - t}{s} \,. \tag{25}$$

Since $y = F(x)$, where y is a uniform random number in $[0, 1)$ and x is a random number according to the Cauchy-Lorentz distribution, we can easily construct the desired random numbers by $x = F^{-1}(y)$. For the application to the continuous function problem, the parameter t is given by the x_i component and the parameter s is set to *max_step*, because for larger values s also larger moves are possible.

For a closer look at these three move classes, they are visualized in Fig. 9.

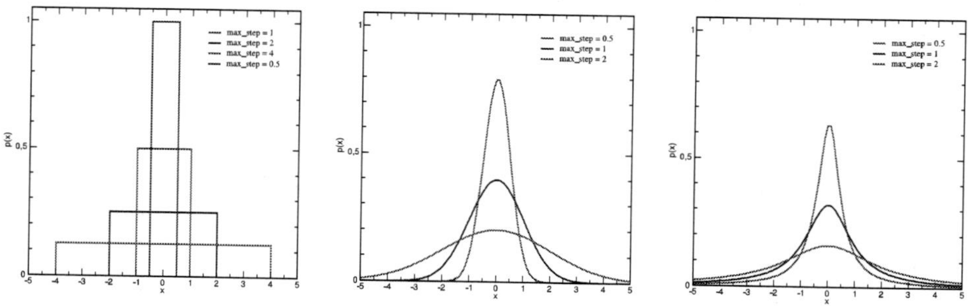

(a) Uniform distribution (b) Gaussian distribution (c) Lorentz distribution

Figure 9. Different probability distributions for move classes in the continuous domain.

It is beneficial to choose the step width adaptive in a way that a fixed percentage of the tried moves is accepted when applying some acceptance strategy as SA. We do this by multiplying it with $1 - \varepsilon$, $\varepsilon \ll 1$, if the acceptance is too low, or with $1 + \varepsilon$, if the acceptance is too high. It seems to be convenient to define a lower bound $c_{\min}$ and an upper bound $c_{\max}$ for the step width additionally.

Basin Hopping: An approach which tries to improve the simple iterative improvement schemes as SA or TA is *Basin Hopping* (BH). BH tries to reduce the original state space $\mathcal{S}$ to a state space $\mathcal{S}^\star \subset \mathcal{S}$, consisting only of the local minima of the original state space. In each iteration a nearby local minimum to the current state is determined and with the rules as given in Equ. (16) or (17) it is accepted as a new state or rejected. In detail, to find a nearby local minimum, one applies a move in a certain neighborhood of some initial size to a state α, getting a different state α' and then one applies local search, e.g. using Equ. (15) to converge to a state β. If $\beta \neq \alpha$ then a criterion $P_{\beta\alpha}$ as e.g. Equ. (16) is applied and β is accepted as new current state or rejected. Otherwise, i.e. if $\beta = \alpha$ after the local search step, the neighborhood has to be increased to hopefully get out of the local optimum with larger jumps.

4.5. Evolutionary Algorithms

Evolutionary Algorithms (EA) are methods which try to mimic the principles of evolution which are variation and selection in combination with a fitness function to solve an optimization problem. In general, three classes of EA are distinguished:

- Evolution Strategies (ES) [11],
- Genetic Algorithms (GA) [38] and
- Evolutionary Programming (EP) [39].

Today it is difficult to clearly allocate each known algorithm to only one of these classes. Major differences are classically the different application of operators and parameter adaption. For an introduction to GA and EP we refer to the literature [38, 39]. ES have been implemented for a comparison to PSO and hence are introduced briefly. In general, the two parameters μ and λ are fixed in advance. The parameter μ is the population size, i.e. individuals are chosen from a population $P^{(i)}$ of u individuals. Then the crossover or mutation operators are applied to produce new individuals, which are then added to an intermediate population. This procedure is repeated, until λ new individuals have been generated. Then for $(\mu+\lambda)$-ES, the new population $P^{(i+1)}$ is chosen from all $\mu+\lambda$ individuals, choosing deterministically the μ best individuals. For (μ,λ)-ES the new population is only chosen from the λ new individuals. This is a good choice especially for dynamic problems, where the fitness landscape changes over time. Very sophisticated self-adaption mechanisms have been developed, which are not introduced here, see e.g. [40].

For our studies here a simple $(\mu+\lambda)$-ES has been applied, choosing $\mu = 30$ and $\lambda = 30$. To optimize functions in the continuous d dimensional space, the move classes from Sec. 4.4. have been applied as mutation operators. Crossover has been realized by different strategies as taking the mid point between two chosen parent states, exchanging coordinates at a randomly chosen dimension between two states and also by classical one point crossover.

Of course there are many more methods which could be compared to PSO, as e. g. *Tabu Search* which makes extensive use of memory, *Model-based Search* or *Extremal Optimization*. A good overview on different techniques is given in [41]. Here we restrict ourselves to the methods as introduced so far.

4.6. Experimental Comparison

The algorithms are applied to a selection of benchmark functions. All results for the selected test functions are summarized in Tab. 8 and 9, where the average solutions for twenty runs are given. In general for at maximum 20,000 objective function evaluations PSO is the superior method. Only for larger running times the other approaches show their benefits and certain algorithms even find better solutions.

The optimization run terminates when the global optimum is approximated by an additive error of 10^{-6} or at most 100,000 function evaluations are done. The algorithms as introduced above are compared to the PSO2 and PSO4 variant, where the results are also determined for 100,000 function evaluations. Again we average the solutions over twenty runs.

Tables 8 and 9 show that the Step function is relatively simple to optimize for each method, at least if the acceptance rule is different from greedy and great deluge. For PSO and BH only the Rastrigin function is difficult to optimize, where PSO performs better for this function. ES perform much better for this function but has weaknesses applied to the Rosenbrock function. The Lorentz moves have several weaknesses but Ensemble-based TA performs very good on most functions except for the Rastrigin function, where interestingly simple SA and TA perform best. An important result is that for each of the applied test functions we found an algorithm which finds the optimum or comes very close to it. This strongly suggests to use hybrid methods combining different approaches in one algorithm, which has been tested as well but is not reported here in detail. Because this chapter mainly focuses on PSO we should especially point out that at least the PSO variants PSO2 and PSO4 are very competitive. All other combinations of a specific method and a specific move class and also ES show at least for one benchmark function more significant weaknesses than

Table 8. Comparison of the average results over 20 runs for different algorithms on continuous benchmark functions using different move classes in $d = 10$

	Algorithm	uniform	Gauss	Lorentz	Gauss adaptive	Lorentz adaptive	BH
Ackley	Greedy	19.688993	19.620313	3.845109	19.499258	15.149652	0.000055
	Simulated Annealing	6.422984	1.478764	2.213738	19.882887	20.509993	0.000053
	Threshold Accepting	15.392034	1.366079	2.142136	19.857492	20.437377	0.000055
	Great Deluge Algo.	19.663482	19.750241	18.193347	17.552375	19.698334	0.000078
	Ensemble Based SA	14.906920	3.478268	2.352309	0.000043	0.000057	0.000054
	Ensemble Based TA	18.142196	5.211392	2.210025	0.000042	0.000038	0.000059
Griewangk	Greedy	0.591554	0.731671	0.939084	6.860117	0.413350	0.010174
	Simulated Annealing	0.673403	0.737713	0.959845	0.305347	0.146674	0.008945
	Threshold Accepting	0.633474	0.781158	0.910945	0.775209	0.117154	0.007473
	Great Deluge Algo.	0.871103	0.975350	1.217548	278.61054	200.78203	0.017534
	Ensemble Based SA	0.665590	0.760320	0.941715	0.071385	0.080507	0.008686
	Ensemble Based TA	0.654103	0.667472	0.924869	0.107090	0.097247	0.007209
Rastrigin	Greedy	81.398649	94.143229	9.425358	77.606349	46.265433	9.452164
	Simulated Annealing	42.998451	31.825698	8.585823	35.719338	1.392988	8.755690
	Threshold Accepting	40.136802	33.395000	8.266775	25.665389	1.282169	9.352667
	Great Deluge Algo.	33.231085	40.755985	18.027384	56.539263	3.942567	11.541578
	Ensemble Based SA	37.018373	37.530228	10.938434	31.739151	8.457204	10.347622
	Ensemble Based TA	47.171130	35.981680	10.098198	25.072942	5.074334	9.452178
Rosenbrock	Greedy	3.005618	2.906340	5.923406	2.745259	0.411125	0.129283
	Simulated Annealing	0.962249	1.654565	4.276056	1.205173	0.015108	0.102862
	Threshold Accepting	2.243364	1.913408	4.089847	9.085765	0.007784	0.081947
	Great Deluge Algo.	3352.0658	4008.6484	6.192481	2174.2755	1840.8896	0.116360
	Ensemble Based SA	2.803984	2.438594	4.213036	1.227660	1.242303	0.065023
	Ensemble Based TA	3.237021	3.232502	3.580854	1.860593	1.084215	0.020712
Step	Greedy	-10.7	-7.1	-60.0	-7.9	-46.4	-60.0
	Simulated Annealing	-60.0	-60.0	-60.0	-59.6	-60.0	-60.0
	Threshold Accepting	-60.0	-60.0	-60.0	-59.9	-60.0	-60.0
	Great Deluge Algo.	-28.8	-50.6	-60.0	-53.6	-60.0	-60.0
	Ensemble Based SA	-60.0	-60.0	-60.0	-60.0	-60.0	-60.0
	Ensemble Based TA	-60.0	-60.0	-60.0	-60.0	-60.0	-60.0

PSO.

This marks the end of the section and also the end of the analysis and comparison of different PSO methods by artificial benchmark functions. Our analysis showed that PSO is very competitive for these instances. In the last section

Table 9. Performance of Evolution Strategies compared to other approaches

Algorithm	Evolution Strategy	Best results of Tab. 8	PSO2	PSO4
Ackley	0.004966	0.000038	0.000000	0.000000
Griewangk	0.301860	0.007209	0.046014	0.051536
Rastrigin	0.004522	1.282169	3.631600	2.313972
Rosenbrock	6.968686	0.007784	2.665125	2.793544
Step	-60.0	-60.0	-60.0	-60.0

of this chapter we focus on the application of PSO to real world problems for the exemplary case of optimizing a hub and spoke inventory system by simulation-based optimization.

5. Simulation-Based Optimization of a Hub and Spoke Inventory System

Often real-world problems in economy and engineering can not be solved strong mathematically and often no specific algorithms are known for these problems. Hence, one has to apply optimization heuristics to obtain good solutions. In this context often simulation-based optimization has to be used since the real-world problem itself cannot be described by an easy and exact model but rather has to be simulated.

As one example, in the following the optimization of a hub and spoke inventory system is presented. Such systems are of special interest because in the age of globalization and increasing competition more and more companies are faced with the problem of how to control the flow of goods from the production locations to the points of demand, i.e. defining optimal transshipment resources and using them efficiently to serve demands with a given number of transportation units [42, 43]. Hub and spoke systems, which consist of one central hub, which is connected to all spokes, arranged similarly to a bicycle wheel, have several benefits including the fact that for a network of d spokes, only d routes are necessary to connect all nodes and this generally leads to a more efficient use of scare transportation resources. Thus, the model is commonly used in industry, in particular in transport, telecommunications and freight, as well as in distributed computing. Additionally, complicated operations can be consolidated at the hub, rather than maintained separately at each node and therefore the spokes are simpler and new spokes can be connected rapidly [44].

Inventory systems with hub and spoke structure have been investigated before [42, 45]. In this section results are shown which in a given sense combine approaches from inventory theory and logistics to investigate the following problem [46, 47]:

- The system consists of one single warehouse, the *hub*, and a number of d retailers, the *spokes*.
- The *hub* owns an infinite amount of one specific product, which can be ordered by the retailers, and a finite number of T transport units. Transportation orders from retailers, which cannot be served immediately due to a stock-out or no available transportation units, are queued in an order queue with given capacity. Ordering and storing of products as well as holding and leasing of transportation units and realizing transshipment generates costs.
  Thus, the following parameters should be concretized: number of products, storage capacities for all products, capacity of the order queue for waiting transshipment orders, number of own transportation units, and cost and gain structure with its parameters.
- The systems allows the realization of different classes of *transportation units* (trucks, pickups, ships, etc.). The parameters of each class are: the number of transportation units in this class, the loading capacity Q of a single transportation unit, the transportation time (velocity), which may be deterministic or random, and the cost structure with its parameters.
- Each *retailer* has a storage for the products with the current level x and there is a stochastic demand of these product units. Depending on the ordering policy each retailer can order new products from the hub. Arriving demand is handled according to an acceptance-rejection policy.
  Thus, the following parameters describe a retailer: distances from the warehouse, the number of products, the client classes with parameters of their demand process, the storage capacities for the products, and the cost and gain structure including holding costs for inventory on hand as well as waiting and rejection costs for backlogged or rejected demand.

The whole system is realized in a simulation tool written by Hochmuth [48], which we will use as an integrated part in the optimization process. A screen shot of this system is shown in Fig. 10. Concretizing all the introduced parameters and choosing the variables and corresponding criteria to be optimized allows to model and optimize a large variety of real-world hub and spoke inventory systems.

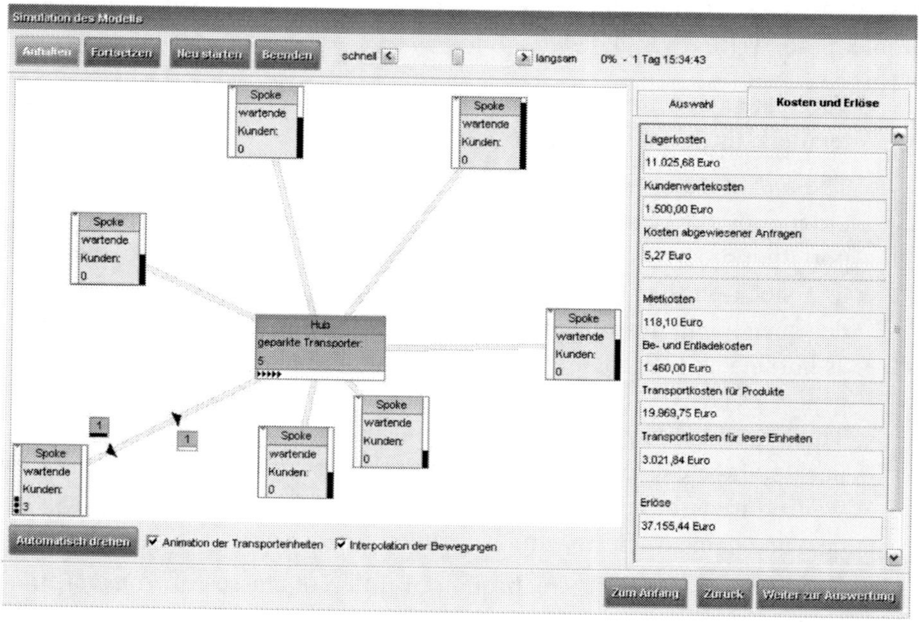

Figure 10. Screenshot of simulation for of hub and spoke inventory system.

The optimization problem is to define an allocation policy of free transportation resources to retailer orders for the warehouse, and for the retailers such ordering policies that minimize the long-run average cost in the system. We basically have the choice between two different options for the ordering policy:

- The (s,S)-*ordering policy* is the restocking to $x = S$ product units when the inventory level of the retailer has fallen below the reorder point s.
- The (s,nQ)-*ordering policy* is the ordering of an integer multiple n of the capacity Q of product units when the inventory level of the retailer has fallen below the reorder point s.

For our inventory system the solution α_i of a swarm individual i encodes the variables for the ordering policy, i.e. the reorder levels s_l for all retailers

l and, if necessary, the number of transportation units t_k for each class k as a real-valued vector. Since a feasible solution can only contain integer numbers, we simply round the real-valued entries of $\mathbf{r}_i$ to obtain the input variables for the simulation. In our view this approach leads to the most natural movement of the swarm individuals in an integer solution space and, further, allows the unchanged use of the common PSO variants originally designed for continuous optimization problems.

Another problem we faced for these systems is the stochastic noise in the objective function, which adds non-static local maxima and minima to the solution space. To reduce this effect one has to choose a sufficiently large time period for the simulation, and it is also advisable to average over several optimization runs. Further, we found by extensive tests that it is reasonable to initially start with relatively low reordering levels since the simulation of the system shows strong fluctuations of the total profit for higher reorder levels s due to fluctuations of the storage costs. In particular, for our test cases the initial values for the reorder level are chosen randomly in a region of the solution space with $s < 20$ and the initial velocities v_i are set to zero.

Since an evaluation of the objective function during the optimization in the PSO-approach requires the complete simulation of the whole system, we are dealing with a very time-consuming optimization problem. For instance a single simulation of the examples discussed in the following can take up to five seconds. Therefore, it is desirable to use optimization algorithms with powerful move classes instead of simple ones in order to minimize the number of such objective function evaluations. One example of such a move class is PSO2, which was identified in Sec. 2.2.2. to converge very fast to an optimum.

System Setup for Hub and Spoke System: In the following we show the optimization results for two test cases. The first deals with a system of identical retailers and the second case with retailers varying either in their distances to the hub or in their demand structure. However, the general setup for all model elements is defined at first. For the warehouse we assume

- a FIFO service policy, and
- an allocation policy with priority sequence according to maximum velocity, maximum loading, minimum empty units, and minimum transport costs.

The retailers are specified by

- 200 km distance from the warehouse,
- a demand with exponentially distributed inter-arrival times with average value of 1 hour and demand of 1 product unit,
- impatient clients with zero waiting time (i.e. the lost-sales case),
- a storage capacity of 500 product units,
- cost parameters: inventory costs 50€ per day and product unit, rejection costs 50€ per rejected client and reward 200€ per satisfied demand unit, and
- all retailers use a (s, nQ) ordering policy with reorder batch size Q, which is equal to the capacity of the used transportation units.

For the transportation units we distinguish between two different classes:

- fast, small transportation units (delivery vans) and not so fast but larger transportation units (trucks),
- their average velocity is equal to 80 km/h and 50 km/h, respectively,
- the capacities are 1 product unit or 4 product units, respectively, and
- the corresponding costs for amortization 50€ or 100€ per day a transportation unit is used, 0.10€ or 0.20€ transportation costs per km and transported product unit, 10€ or 30€ loading/unloading costs, respectively.

The optimization criterion is maximization of the total profit over the planning period, which is chosen to equal 3 years. For the evaluation of the objective function for each solution we average over 5 independent simulation runs with the same parameters in order to minimize fluctuations. Further, the starting inventory is 10 product units for each retailer.

By a few pre-considerations, we obtain that the average demand per day is about 120 product units. One fast transport unit can deliver at maximum 4.8 product units per day, whereas a slow transport unit can deliver 12 product units. This corresponds to either the usage of 25 fast or 10 slow transport units a day and also motivates the first example, where we will investigate the influence of different numbers of available transportation units.

Example – 5 identical retailers: In this example we use the setup for the model components as specified above and the numbers of transportation units are chosen according to the following cases:

a) 20 transportation units with velocity 80 km/h and capacity 1 product unit,
b) 8 transportation units with velocity 50 km/h and capacity 4 product units,
c) 10 fast transportation units and 4 slow transportation units with capacities according to a) and b), and
d) optimization of the number of fast and slow transportation units.

Additionally, we consider two setups for the reorder level s. Because we assume identical retailers, we first choose identical s for all retailers. After this we optimize s individually for each retailer. The results are expected to be nearly identical for both cases and are shown in the format $\mathbf{x}^\star = (\mathbf{s}^\star, \mathbf{t}^\star)$ with the optimal reorder levels $\mathbf{s}^\star = (s_1^\star, \ldots, s_5^\star)$ and the number of transportation units $\mathbf{t}^\star = (t_{\text{fast}}^\star, t_{\text{slow}}^\star)$ (cp. Table 10).

Table 10. Results for a hub and spoke inventory system *with 5identical retailers*

Case		Ordering Strategy		Total profit in €
identical reorder levels s for each retailer	a	$\mathbf{s}^\star = 6$		14.425.646
	b	$\mathbf{s}^\star = 7$		13.487.500
	c	$\mathbf{s}^\star = 18$		13.058.093
	d	$\mathbf{s}^\star = 7$	$\mathbf{t}^\star = (33,0)$	19.014.137
individual reorder levels s for each retailer	a	$\mathbf{s}^\star = (5,5,6,5,6)$		14.407.362
	b	$\mathbf{s}^\star = (3,11,11,11,3)$		13.770.108
	c	$\mathbf{s}^\star = (18,18,18,18,18)$		13.043.158
	d	$\mathbf{s}^\star = (8,9,7,7,6)$	$\mathbf{t}^\star = (32,0)$	18.917.484

The search finished for identical s in less than 10 iterations for PSO with 10 concurrent individuals. The results are shown in Tab. 10 and are identical or very close for this case, surely caused by the very low number of degrees of freedom, which are either only 1 or 3. For the case in which the reorder levels s_l are optimized for each retailer l individually, we expect to obtain similar results. This is more or less satisfied. Further, in case b), i.e. 8 large transportation units, there are noticeable fluctuations in the reorder level, which seems to be caused by only small differences between different reordering strategies in this area of the solution space. The solutions are also very close to that of identical s, which shows that PSO is able to adequately optimize this inventory systems with more degrees of freedom.

When optimizing also the number of transportation units, we can increase the solution quality considerably. The results show that the operation of fast transportation units yields in this example much better results, and hence, the optimized solution only uses fast ones, where the optimal number of 32 or 32 transportation units is greater than necessary for the satisfaction of the average daily demand, given by 25 fast TUs. The reason might be that it is optimal for the system to have enough products in the storage in order to avoid rejection costs.

Example – 5 different retailers, 2 types of TUs: In this example we use the setup for the model components as specified above and optimize the reorder levels of the retailers as well as the number of fast and slow TUs for the following cases:

a) *Different demand in the retailers:* The retailers demand is given by an exponential distribution with average inter-arrival times $ET_1 = 1.5\text{h}$, $ET_2 = 1.2\text{h}$, $ET_3 = 1\text{h}$, $ET_4 = 6/7\text{h}$ and $ET_5 = 3/4\text{h}$. This results in an average daily demand of 16, 20, 24, 28 and 32 product units respectively, which also sums up to the former 120 product units.

b) *Different distances from the warehouse:* Again this is based on Example 1, but with different retailer-warehouse distances. We consider two variants, where variant 2 has the same average distances as Example 1:
Variant 1: $d_1 = 50\text{km}$, $d_2 = 100\text{km}$, $d_3 = 200\text{km}$, $d_4 = 300\text{km}$, $d_5 = 400\text{km}$
Variant 2: $d_1 = 100\text{km}$, $d_2 = 150\text{km}$, $d_3 = 200\text{km}$, $d_4 = 250\text{km}$, $d_5 = 300\text{km}$

For the setup in a) we expect a general tendency towards an increase of the reorder levels from retailer 1 to 5 since the average daily demand also increases in this direction. The results are summarized in Tab. 11. For case a) we clearly can see the expected increase in the reorder levels. Further, the results are almost identical to the total profit of about 19 Mio. € obtained in Example 1. However, in the PSO solution the number of TUs is significantly increased compared to Example 1 and for both algorithms also the usage of a few slow TUs is found to be optimal. The reason for the different numbers of TUs is probably again the relatively flat shape of the solution space.

In case b) we again obtain for both variants the expected increase in the reorder levels. This has been expected because the retailers with larger distance have to hold more products in the storage to compensate the longer delivery times. Since variant 2 has a shorter average distance than variant 1, the total profit of variant 2 is slightly better. Further, comparing the results of variant 2 with Example 1, which has the same average setup, we obtain very similar results for the total profit as expected.

Table 11. Results for a hub and spoke inventory system *with 5 different retailers*

Case	Ordering Strategy		Total profit in €
different demand	$s^\star = (5,6,7,8,8)$	$t^\star = (41,5)$	19.032.411
different distances variant 1	$s^\star = (6,6,8,9,10)$	$t^\star = (34,0)$	18.847.544
different distances variant 2	$s^\star = (5,8,7,8,9)$	$t^\star = (31,0)$	18.895.547

6. Conclusion

In this chapter we compared different approaches to particle swarm optimization. First four different variants have been introduced: the relatively simple classical approach by Kennedy and Eberhart which needs only two random numbers per iteration and individual in the swarm, a more involved version using matrices of random numbers and an optimized parameter setup by Clerc and Kennedy, a computational expensive approach applying matrix rotations and charged PSO, a version which comes in different flavors, where a repulsive force term gets introduced.

Besides the different equations of motion, which is the major ingredient of a PSO algorithm, many other characteristics can be modified or constrained, i.e. the initial particle velocities, the feasible velocities during the algorithm execution, the weight factor and the neighborhood of the particles. Also different boundary conditions for the handling of the boundaries of the search space are applicable.

A first broadly based comparison of the four PSO variants for fourteen standard benchmark functions in ten dimensions for 20,000 function evaluations showed, that the PSO1 variant is not competitive compared with the other three approaches, but the results were inconclusive to find a superior method within this group. But a disadvantage of PSO3 was already here a much higher computation time compared to PSO2 and PSO4.

A closer look at the convergence behavior of the different methods for a subset of three hard test functions was the next step. First, the swarm size has been varied from 5 up to 60 individuals. Where the performance gain by using larger swarms is high for the simple PSO1 approach, the optimum for the other three variants is given with about 30 to 40 individuals in the swarm. Also the dimension of the problems has been varied between 2 and 50. This showed again that the PSO1 variant may not be the best choice. Comparing the other three methods applied to higher-dimensional problems, PSO4 performs better than the other two approaches, except PSO3 for the Rosenbrock function. But the advantage is moderate and PSO4 achieved its results in a much lower running time.

In further investigations, some of the heuristics and boundary conditions turned out to be contra-productive as e.g. initial velocities for PSO2 to PSO4 or minimum velocities. Also maximum velocities had only for PSO1 a positive influence. The same holds basically for absorbing boundary conditions (few improvements have been obtained for PSO4). Reflecting boundary conditions were contra-productive for all variants.

In further comparative studies, other algorithms as Simulated Annealing, Threshold Accepting, Basin Hopping for different move classes, and Evolution Strategies were compared to PSO. The results show that PSO is competitive, especially for relatively low numbers of function evaluations. For higher numbers of function evaluations, other methods perform for some functions equally well or better. Again, the results are inconclusive - there is no method, which dominates all other methods.

Finally, a study of the application of PSO to the simulation-based optimization of hub and spoke inventory systems showed that PSO can not only be applied successfully to benchmark functions but also to quite complex practical problems.

References

[1] E. Schöneburg, F. Heinzmann, and S. Feddersen. *Genetische Algorithmen und Evolutionsstrategien - Eine Einführung in Theorie und Praxis der simulierten Evolution.* Addison-Wesley, Bonn, 1994. ISBN 3-89319-493-2.

[2] J. D. Farmer, N. H. Packard, and A. S. Perelson. The Immune System, Adaptation and Machine Learning. *Physica D: Nonlinear Phenomena*, **22**(1-3):187–204, October-November 1986.

[3] A. Colorni, M. Dorigo, and V. Maniezzo. Distributed Optimization by Ant Colonies. In F. J. Varela and P. Bourgine, editors, *Towards a Practice of Autonomous Systems: Proceedings of the First European Conference on Artificial Life*, pages 134–142. MIT Press, 1992.

[4] J. Kennedy and R. Eberhart. Particle Swarm Optimization. *Proceedings of IEEE International Conference on Neural Networks*, **4**:1942–1948, November-December 1995.

[5] K.-P. Wang, L. Huang, C.-G. Zhou, and W. Pang. Particle Swarm Optimization for Traveling Salesman Problem. *Proceedings of the 2nd International Conference on Machine Learning and Cybernetics*, **3**:1583–1585, November 2003.

[6] S.-C. Chu, Y.-T. Chen, and J.-H Ho. Timetable Scheduling Using Particle Swarm Optimization. *First International Conference on Innovative Computing, Information and Control*, **3**:324–327, August 2006.

[7] S. Kirkpatrick, C. D. Gelatt Jr., and M. P. Vecchi. Optimization by Simulated Annealing. *Science*, **220**(4598):671–680, May 1983.

[8] G. Dueck and T. Scheuer. Threshold Accepting: a General Purpose Optimization Algorithm Appearing Superior to Simulated Annealing. *Journal of Computational Physics*, **90**(1):161–175, September 1990.

[9] G. Ruppeiner, J. M. Pedersen, and P. Salamon. Ensemble Approach to Simulated Annealing. *Jounal de Physique*, (**1**):455–470, January 1991.

[10] P. Salamon, P. Sibani, and R. Frost. *Facts, Conjectures and Improvements for Simulated Annealing*. SIAM, Society for Industrial and Applied Mathematics, Philadelphia, 2002. ISBN 0-898-71508-3.

[11] I. Rechenberg. *Evolutionsstrategie '94*. Frommann Holzboog, Stuttgart Bad Canstatt, 1994. ISBN 978-3-772-81642-0.

[12] Y. Shi and R. Eberhart. A Modified Particle Swarm Optimizer. *'IEEE World Congress on Computational Intelligence' - Proceedings of the 1998 IEEE International Conference on Evolutionary Computation*, pages 69–73, May 1998.

[13] J. S. Vesterstrom, J. Riget, and T. Krink. Division of Labor in Particle Swarm Optimisation. *Proceedings of the Evolutionary Computation*, 2:1570–1575, August 2002.

[14] S. Yang, M. Wang, and L. Jiao. A Quantum Particle Swarm Optimization. *Congress on Evolutionary Computation*, **1**:320–324, June 2004.

[15] G. Ciuprina, D. Ioan, and I. Munteanu. Use of Intelligent-Particle Swarm Optimization in Electromagnetics. *IEEE Transactions on Magnetics*, **38**(2):1037–1040, March 2002.

[16] D. N. Wilke. *Analysis of the Particle Swarm Optimization Algorithm*. Master Thesis, University of Pretoria, 2005.

[17] A. Carlisle and G. Dozier. An off-the-Shelf PSO. *Proceedings of the Workshop on Particle Swarm Optimization*, **1**:1–6, April 2001.

[18] C. Molery and C. Van Loanz. Nineteen Dubious Ways to Compute the Exponential of a Matrix, Twenty-Five Years Later. *SIAM Review*, **45**(1):3–49, March 2003.

[19] M. Clerc and J. Kennedy. The Particle Swarm - Explosion, Stability, and Convergence in a Multidimensional Complex Space. *IEEE Transactions on Evolutionary Computation*, 6(1):58–73, February 2002.

[20] O. Urfalioglu. Robust Estimation of Camera Rotation,Translation and Focal Length at High Outlier Rates. *Proceedings of the First Canadian Conference on Computer and Robot Vision*, pages 464–471, May 2004.

[21] M. Blackwell. Particle Swarms and Population Diversity. *Soft Computing – A Fusion of Foundations, Methodologies and Applications*, **9**(11):793–802, May 2005.

[22] S. Thiem, J. Lässig, and P. Köchel. Merging Groups for the Exploration of Complex State Spaces in the CPSO Approach. In *ANTS '08: Proceedings of the 6th international conference on Ant Colony Optimization and Swarm Intelligence*, volume 5217/2008, pages 355–362. Springer, 2008.

[23] W. M. Spears, D. F. Spears, J. C. Hamann, and R. Heil. Distributed, Physics-Based Control of Swarms of Vehicles. *Autonomous Robots*, **17**(2-3):137–162, September 2004.

[24] R. Eberhart and J. Kennedy. A New Optimizer using Particle Swarm Theory. *Proceedings of the Sixth International Symposium on Micro Machine and Human Science*, pages 39–43, October 1995.

[25] Y.-L. Zheng, L.-H. Ma, L.-Y. Zhang, and J.-X. Qian. On the Convergence Analysis and Parameter Selection in Particle Swarm Optimization. *International Conference on Machine Learning and Cybernetics*, **3**:1802–1807, November 2003.

[26] A. S. Mohais. *Random Dynamic Neighborhood Structures in Particle Swarm Optimization*. PhD Thesis, The University of the West Indies, 2007.

[27] H. Gould, J. Tobochnik, and W. Christian. *An Introduction to Computer Simulation Methods – Applications to Physical Systems*. Addison-Wesley, San Francisco, 3rd edition, 2006. ISBN 0-805-37758-1.

[28] J. Lässig and S. Thiem. An Integrated Framework for the Realtime Investigation of State Space Exploration. *Proceedinfs of the 5th International Conference on Computational Intelligence*, February 2008, pp. 24-29.

[29] J. Robinson and Y. Rahmat Samii. Particle Swarm Optimization in Electromagnetics. *IEEE Transactions on Antennas and Propagation*, **52**(2):397–407, February 2004.

[30] A. Franz and K. H. Hoffmann. Optimal Annealing Schedules for a Modified Tsallis Statistics. *Journal of Computational Physics*, **176**(1):196–204, February 2002.

[31] F. Heilmann. *The State Space of Complex Systems*. Dissertation, Chemnitz University of Technology, 2005.

[32] A.-R. Hedar. Test Functions for Unconstrained Global Optimization. http://www-optima.amp.i.kyoto-u.ac.jp/member/student/hedar/Hedar_files/TestGO_files/Page364.htm, 2009. [Online: accessed 17-October-2009].

[33] The TRACER Project. The Problem Site. http://tracer.lcc.uma.es/problems/index.html, 2004. [Online: accessed 17-October-2009].

[34] C. Blum and A. Roli. Metaheuristics in Combinatorial Optimization: Overview and Conceptual Comparison. *ACM Computing Surveys*, **35**(3):305–310, September 2003.

[35] G. Dueck. New Optimization Heuristics: The Great Deluge Algorithm and the Record-to-Record Travel. *Journal of Computational Physics*, **104**(1):86–92, January 1993.

[36] A. Franz, K. H. Hoffmann, and P. Salamon. Best Possible Strategy for Finding Ground States. *Physical Review Letters*, **86**(23):5219–5222, June 2001.

[37] G. E. P. Box and M. E. Muller. A Note on the Generation of Random Normal Deviates. *The Annals of Mathematical Statistics*, **29**(2):610–611, June 1958.

[38] J. H. Holland. *Adaptation in Natural and Artificial Systems*. University of Michigan Press, Ann Arbor, 1975. ISBN 0-387-58334-3.

[39] L. Fogel, A. Owens, and M. Walsh. *Artificial Intelligence through Simulated Evolution*. Wiley, New York, 1966.

[40] H.-G-Beyer. *The Theory of Evolution Strategies*. Springer, Berlin, Heidelberg, New York, 2001. ISBN: 3-540-67297-4.

[41] R. Battiti, M. Brunato, and F. Mascia. *Reactive Search and Intelligent Optimization*. Technical Report DIT-07-049, Università di Trento, Dipartimento di Informatica e Telecomunicazioni, 2007.

[42] P. Köchel, S. Kunze, and U. Nieländer. Optimal Control of a Distributed Service System with Moving Resources: Application to the Fleet Sizing and Allocation Problem. *International Journal of Production Economics*, **81-82**(1):443–459, January 2003.

[43] P. Köchel and U. Nieländer. Simulation-based optimisation of multi-echelon inventory systems. *International Journal of Production Economics*, **93-94**(1):505–513, January 2005.

[44] Wikipedia. Spoke-Hub Distribution Paradigm — Wikipedia, The Free Encyclopedia. `http://en.wikipedia.org/w/index.php?title=Spoke-hub_distribution_paradigm&oldid=319981898`, 2009. [Online; accessed 20-November-2009].

[45] G. Zhou and M. Min, H.; Gen. The Balanced Allocation of Customers to Multiple Distribution Centers in the Supply Chain Network: a Genetic Algorithm Approach. *Computers & Operations Research*, **43**(1-2):251–261, July 2002.

[46] P. Köchel. Hub-and-Spoke-Systems: On Optimal Allocations of Transportation Resources in the Hub under Optimal (s,nQ) Ordering Policies at the Spokes. *2. Deutsch-Russischer Logistik-Workshop*, May 2007.

[47] P. Köchel and S. Thiem. Search for Good Policies in a Single-warehouse, Multi-retailer System by Particle Swarm Optimisation. In *15th International Symposium on Inventories*, Budapest, Hungary, August 2008.

[48] C. A. Hochmuth. *Modellierung, Simulation und Visualisierung eines Hub-and-Spoke-Systems*. Term Subject, Chemnitz University of Technology, 2007.

In: Mathematical Theories, Models and Groups ISBN 978-1-61324-250-6
Editor: Marcus J. Grecho

Chapter 2

DETECTING THE COMMUTING PROBABILITY OF THE DERIVED SUBGROUP

A. Erfanian[1,*] ***and F. Russo***[2,†]
[1]Department of Mathematics,
Centre of Excellence in Analysis on
Algebraic Structures,
Ferdowsi University of Mashhad, Mashhad, Iran
[2]Department of Mathematics,
University of Naples Federico II, Naples, Italy

Abstract

There is a long standing line of research, which is devoted to investigate bounds for $|G'|$ when G is an infinite group. This line goes back to a classic result of I. Schur. The present paper deals with the structure of G' when G is a compact group, showing that $|G'|$ can be controlled by the notion of commuting probability.

2000 Mathematics Subject Classification. Primary: 20D60, 20P05; Secondary: 20D08.

Keywords: Commuting probability, compact group, p-elementary abelian groups of rank 2, derived subgroup.

*E-mail address: erfanian@math.um.ac.ir
†E-mail address: francesco.russo@dma.unina.it

1. Introduction

If G is a finite group, the probability that a randomly chosen pair of elements of G commutes is defined to be $\#com(G)/|G|^2$, where $\#com(G)$ is the number of pairs $(x,y) \in G \times G$ with $xy = yx$, G^2 is the product of two copies of G and $|G|^2$ is the order of G^2. Note that this ratio is denoted by $cp(G) = k(G)/|G|$ in [1, 3, 5, 10] where $k(G)$ is the number of the conjugacy classes of G. See also [2, 4, 7]. More precisely, if G is a finite group,

$$cp(G) = \frac{|\{(x_1,x_2) \in G^2\ ;\ x_ix_j = x_jx_i \text{ for all } 1 \leq i,j \leq 2\}|}{|G|^2}.$$

If G is a non-abelian group, then $cp(G) \leq 5/8$; furthermore this bound is achieved if and only if $G/Z(G)$ has order 4, where $Z(G)$ is the center of G. Such a result can be found in [5].

The ratio $cp(G)$ has been extended to a compact group G already in [5, Section 2], defining $cp(G) = (\mu \times \mu)(C)$, where $C = \{(x,y) \in G^2 \mid xy = yx\}$, $f : (x,y) \in G^2 \rightarrow [x,y] \in G$, $C = f^{-1}(1)$ and μ is the normalized Haar measure on G. Note that C is measurable, since it is the anti-image of the closed set $\{1\}$ under the map f which is continuous (see [5, Section 2]). These information and the properties of the Haar measure on G guarantee that $cp(G)$ is well–defined (see also [6, Chapter 2]). Obviously, if G is finite, then it is a compact group with the discrete topology and so the Haar measure on G is the counting measure. Most of the results in [1, 3, 4, 5, 7, 10] can be seen in such a way. We list now our main results. Section 2 will allow us to prove them in Section 3.

Theorem A. *Let G be a non-abelian connected compact group, $Z_0(G)$ be the identity component of $Z(G)$ and $G/Z_0(G)$ be a p-group, where p is a prime. Then the following statements are equivalent:*

(i) *$G/Z_0(G)$ is a p-elementary abelian group of rank 2;*

(ii) *G' is a p-elementary abelian group of rank 2;*

(iii) $cp(G) = \frac{p^2+p-1}{p^3}$.

Theorem B. *Let G be a non-abelian compact group, $sol(G/Z(G))$ be the solvable radical of $G/Z(G)$, $F(G/Z(G))$ be the Fitting subgroup of $G/Z(G)$, d be* *the maximum number of elements in a conjugacy class of G, l be the derived length of $G/Z(G)$, p be a prime and n, m be positive integers.*

(i) *If* $|G/Z(G)| = n$, *then* $d^{-\frac{1}{2}(1+log_2 d)} \leq |G'|^{-1} < cp(F(G/Z(G)))^{\frac{1}{2}}|G/Z(G) : F(G/Z(G))|^{-\frac{1}{2}} \leq |G/Z(G) : F(G/Z(G))|^{-\frac{1}{2}}$.

(ii) *If* $G/Z(G)$ *is soluble of order n, then* $d^{-\frac{1}{2}(1+log_2 d)} \leq |G'|^{-1} < log_2(|G/Z(G) : sol(G/Z(G))|)^{-\frac{1}{3}}$.

(iii) *If* $|G/Z(G)| = n$, *then* $d^{-\frac{1}{2}(1+log_2 d)} \leq |G'|^{-1} < |G/Z(G) : sol(G/Z(G))|^{-\frac{1}{2}}$.

(iv) *If* $G/Z(G)$ *is finite soluble with* $l \geq 4$, *then* $d^{-\frac{1}{2}(1+log_2 d)} \leq |G'|^{-1} < \frac{4l-7}{2^{l+1}}$.

(v) *If* $|G/Z(G)| = p^m$, *then* $p^{-\frac{1}{2}m(m-1)} \leq |G'|^{-1} < \frac{p^l+p^{l-1}-1}{p^{2l-1}}$.

2. Preliminaries

In this section, G is assumed to be a non-abelian compact group (not necessarily finite even uncountable) with normalized Haar measure μ.

Lemma 2.1. *Let $C_G(x)$ be the centralizer of an element x in G. Then*

$$cp(G) = \int_G \mu(C_G(x))d\mu(x),$$

where $\mu(C_G(x)) = \int_G \chi_C(x,y)d\mu(y)$ *and* χ_C *denotes the characteristic map of the set C.*

Proof. See [2, Lemma 3.1]. ♢

Lemma 2.2. *Let H be a closed subgroup of G, n, r be positive integers and p be a prime.*

(i) *If* $|G : H| \geq n$, *then* $\mu(H) \leq \dfrac{1}{n}$.

(ii) *If* $|G : H| \leq n$, *then* $\mu(H) \geq \dfrac{1}{n}$.

(iii) *Assume that* $G/Z(G)$ *is a* p*-group of order* p^r. *An element* x *belongs to* $Z(G)$ *if and only if* $\mu(C_G(x)) > \frac{1}{p}$.

Proof. See [2, Lemmas 3.2, 3.4]. ◇

Lemma 2.3. *Let* r *be a positive integer. If* $G/Z(G)$ *is a* p*-elementary abelian group of rank* r*, then* $cp(G) \leq \frac{p^r+p-1}{p^{r+1}}$*, for every prime* p*. The equality holds when* $r = 2$.

Proof. Assume that $G/Z(G)$ is a p-elementary abelian group of rank r. By Lemma 2.1 and Lemma 2.2, we have

$$\begin{aligned} cp(G) &= \int_G \mu(C_G(x))d\mu(x) = \int_{G-Z(G)} \mu(C_G(x))d\mu(x) + \mu(Z(G)) \\ &\leq \tfrac{1}{p}(\mu(G) - \mu(Z(G))) + \mu(Z(G)) = \tfrac{1}{p}(1 - \tfrac{1}{p^r}) + \tfrac{1}{p^r} = \tfrac{p^r+p-1}{p^{r+1}} \end{aligned}$$

If $r = 2$, then G is the union of p^2 distinct cosets

$$G = Z(G) \cup x_1Z(G) \cup x_2Z(G) \cup \ldots \cup x_{p^2-1}Z(G)$$

and so $1 = \mu(G) = p^2\mu(Z(G))$. Moreover, if $a, b \in x_iZ(G)$, for $1 \leq i \leq p^2 - 1$, then $a = x_iz_1$ and $b = x_iz_2$ for some $z_1, z_2 \in Z(G)$ so that

$$ab = x_iz_1x_iz_2 = x_ix_iz_1z_2 = x_ix_iz_2z_1 = x_iz_2x_iz_1 = ba.$$

Thus, if $a \in x_iZ(G)$, then $C_G(a) = Z(G) \cup aZ(G) \cup a^2Z(G) \cup \ldots \cup a^{p-1}Z(G)$ and so

$$\begin{aligned} \mu(C_G(a)) &= \mu(Z(G)) + \mu(aZ(G)) + \mu(a^2Z(G)) + \ldots + \mu(a^{p-1}Z(G)) \\ &= p\mu(Z(G)) = p(\tfrac{1}{p^2}) = \tfrac{1}{p} \end{aligned}$$

Thus, we have

$$\begin{aligned} cp(G) &= \int_G \mu(C_G(x))d\mu(x) \\ &= \int_{Z(G)} \mu(C_G(x))d\mu(x) + \textstyle\sum_{i=1}^{p^2-1} \int_{x_iZ(G)} \mu(C_G(x))d\mu(x) \\ &= \mu(Z(G)) + \tfrac{1}{p}\textstyle\sum_{i=1}^{p^2-1} \mu(Z(G)) = (\tfrac{1}{p}(p^2-1) + 1)\mu(Z(G)) \\ &= \tfrac{p^2+p-1}{p^3}. \ \diamond \end{aligned}$$

Proposition 2.4. *Let N be a closed normal subgroup of G. Then*

$$cp(G) \leq cp(G/N).$$

In particular, if $N \cap G' = 1$, then the equality holds.

Proof. Let λ, μ and ν be the corresponding Haar measure of N, G and G/N respectively. Let $x \in G$, $y \in N$ and $xN \in G/N$. The integral properties of the Haar measure on G allow us to write

$$\int_G \mu(C_G(x))du(x) = \int_{G/N} \left(\int_N \mu(C_G(xy))d\lambda(y) \right) d\nu(xN).$$

Since ν is a Haar measure on G/N, ν acts on G/N as μ on G modulo N so that $\mu(C_G(xy)N) = \nu(C_G(xy)N/N)$. But in general $C_G(xy)N/N \leq C_{G/N}(xN)$, so that $\nu(C_G(xy)N/N) \leq \nu(C_{G/N}(xN))$ being ν monotone.

Then, $\mu(C_G(xy)N) = \nu(C_G(xy)N/N) \leq \nu(C_{G/N}(xN))$ and, from Lemma 2.1, we have

$$\begin{aligned}
cp(G) &= (\mu \times \mu)(C) \\
&= \textstyle\int_G \mu(C_G(x))d\mu(x) = \int_{G/N} (\int_N \mu(C_G(xy))d\lambda(y))\, d\nu(xN) \\
&\leq \textstyle\int_{G/N} (\int_N u(C_G(xy)N)d\lambda(y))\, d\nu(xN) \\
&\leq \textstyle\int_{G/N} (\int_N \nu(C_{G/N}(xN))d\lambda(y))\, d\nu(xN) \\
&= \textstyle\int_{G/N} \nu(C_{G/N}(xN)) (\int_N d\lambda(y))\, d\nu(xN) \\
&= \textstyle\int_{G/N} \nu(C_{G/N}(xN))d\nu(xN) = cp(G/N).
\end{aligned}$$

In particular, if $N \cap G' = 1$, then $C_G(xy) = C_G(xy)N$ and so $\mu(C_G(xy)) = \mu(C_G(xy)N)$. Furthermore, $\mu(C_G(xy)N) = \nu((C_G(xy)N)/N) = \nu(C_{G/N}(xN))$. So, the equality holds. ◇

Recall from [6] that G_0 denotes the *identity component* of G. In particular, $Z_0(G)$ denotes the identity component of $Z(G)$.

Lemma 2.5. *If G is connected, then $\mu(G') = \mu(G/Z_0(G))$.*

Proof. From [6, Theorem 9.24 (ii)], $G = Z_0(G)G'$ and $Z_0(G) \cap G'$ is totally disconnected. We conclude that

$$\mu(G) = \mu(Z_0(G)G') = \mu(Z_0(G)) + \mu(G') - \mu(Z_0(G) \cap G')$$

Since $Z_0(G) \cap G'$ is totally disconnected, $\mu(Z_0(G) \cap G') = 0$, and so

$$\mu(G') = \mu(G) - \mu(Z_0(G)) = \mu(G/Z_0(G)). \diamondsuit$$

The proof of Lemma 2.5 uses [6, Theorem 9.24 (ii)] which is a fundamental result in the Theory of Compact Groups. Moreover it allows us to have a precise control of the measure of G' as the following remark shows.

Remark 2.6. From [6, Theorem 9.24 (ii)], if we have a non-abelian compact group G, then there exists a family $\{S_j : j \in J\}$ of simple connected compact Lie groups and a totally disconnected central subgroup D of $Z_0(G) \times \prod_{j \in J} S_j$ such that $G \cong \frac{Z_0(G) \times \prod_{j \in J} S_j}{D}$. Since D is totally disconnected, $\mu(D) = 0$, and so $\mu(G)$ is equal to

$$\mu\Big(\frac{Z_0(G) \times \prod_{j \in J} S_j}{D}\Big) = (\mu(Z_0(G)) + \mu(\prod_{j \in J} S_j)) - \mu(D) = \mu(Z_0(G)) + \mu(\prod_{j \in J} S_j).$$

By Lemma 2.5, $\mu(G') = \mu(G/Z_0(G)) = \mu(\prod_{j \in J} S_j)$. $\diamondsuit$

The following lemma adapts [4, Lemma 2 (vi)].

Lemma 2.7. *Let G be a non-abelian compact group with $|G : Z(G)| = n$, where n is a positive integer. Then $|G'|^{-1} < cp(G)$.*

Proof. A famous bound of Wiegold (see [8, p.102 (2)]) shows that, if $|G : Z(G)|$ is finite, then $|G'|$ is finite as well. Now, we can easily observe that the length of every conjugacy class is bounded above by the order of derived subgroup G' for every element $x \in G$. This means $|G : C_G(x)| \leq |G'|$ for all $x \in G$. By Lemma 2.2, $\mu(C_G(x)) \geq |G'|^{-1}$ for each element $x \in G$. Moreover, if $|G : Z(G)| = n$, then we may write G as the union of n distinct cosets

$$G = Z(G) \cup x_1 Z(G) \cup x_2 Z(G) \cup \ldots \cup x_{n-1} Z(G)$$

and so $\mu(Z(G)) = 1/n$. Thus we will have

$$cp(G) = \int_G \mu(C_G(x))d\mu(x) = \int_{Z(G)} \mu(C_G(x))d\mu(x) + \int_{x_1Z(G)} \mu(C_G(x))d\mu(x) +$$

$$\cdots + \int_{x_{n-1}Z(G)} \mu(C_G(x))d\mu(x) = \mu(Z(G)) + \Sigma_{i=1}^{n-1} \int_{x_iZ(G)} \mu(C_G(x))d\mu(x)$$

$$\geq \frac{1}{n} + \Sigma_{i=1}^{n-1} \int_{x_iZ(G)} |G'|^{-1} d\mu(x) > \frac{1}{n}|G'|^{-1} + \frac{n-1}{n}|G'|^{-1} = |G'|^{-1}. \diamond$$

3. Proofs of Theorems A and B

This Section contains our main results with some instructive examples.

Proof of Theorem A. (i)$\Rightarrow$(ii) From [6, Theorem 9.24], $G = Z_0(G)G'$ so that G' is isomorphic as compact group to $G/Z_0(G)$. Now the property to be a p-elementary abelian group of rank 2 is invariant under isomorphisms of compact groups. Then the result follows.

(ii)$\Rightarrow$(iii). Again from [6, Theorem 9.24] we have that G' is isomorphic to $G/Z_0(G)$ and so $G/Z_0(G)$ is a p-elementary abelian group of rank 2. Since $Z_0(G) \leq Z(G)$ and the class of p-elementary abelian groups is closed with respect to forming subgroups, images and extensions of its members (see [8]), we conclude that $G/Z(G)$ is a p-elementary abelian group of rank 2. Now Lemma 2.3 gives the required bound.

(iii)$\Rightarrow$(i). Assume that $cp(G) = \frac{p^2+p-1}{p^3}$ and $G/Z_0(G)$ is not a p-elementary abelian group of rank 2. By assumption $G/Z_0(G)$ is a p-group. So, if $G/Z_0(G)$ has order 1 or p, then it is cyclic. Since $Z_0(G) \leq Z(G)$, also $G/Z(G)$ is cyclic. It follows that G is abelian and there is a contradiction. Thus $|G : Z_0(G)| \geq p^2$. If $|G : Z_0(G)| = p^2$, then $G/Z_0(G)$ is an abelian of order p^2 and it is either cyclic of order p^2 or a p-elementary abelian group of rank 2. In the first case we obtain again a contradiction and in the second case we finish. Now suppose that $|G:Z_0(G)|>p^2$. Using [6, Theorem 9.24 (ii)] and Lemma 2.5, we have

$$cp(G)=\int_G \mu(C_G(x))d\mu(x)=\int_{Z_0(G)G'}\mu(C_G(x))d\mu(x)$$

$$=\int_{Z_0(G)}\mu(C_G(x))d\mu(x)+\int_{G'}\mu(C_G(x))d\mu(x)-\int_{G'\cap Z_0(G)}\mu(C_G(x))d\mu(x)$$

$$=\int_{Z_0(G)}\mu(C_G(x))d\mu(x)+\int_{G'}\mu(C_G(x))d\mu(x)$$

$$=\int_{Z_0(G)}\mu(C_G(x))d\mu(x)+\int_{G'\setminus Z_0(G)}\mu(C_G(x))d\mu(x)$$

$$\leq \mu(Z_0(G))+(\mu(G')-\mu(Z_0(G)))=\mu(G')=\mu(G/Z_0(G)).$$

But $Z_0(G)$ is a closed normal subgroup of G with $|G:Z_0(G)|>p^2$, then Lemma 2.2 implies $\mu(G/Z_0(G))<\frac{1}{p^2}$. Now the relation $\frac{p^2+p-1}{p^3}<\frac{1}{p^2}$ gives a contradiction and the result follows. ◇

Proof of Theorem B. The finiteness of $G/Z(G)$ implies the finiteness of G'

by a famous Schur's Lemma (see [8, Theorem 4.12]), so there are no problems to consider the maximum number of elements in a conjugacy class of G [8, Theorem 4.35].

Lemma 2.7, combined with [4, Theorem 4 (ii)] and Proposition 2.4, implies

$$|G'|^{-1}<cp(G)<cp(G/Z(G))\leq$$

$$cp(F(G/Z(G)))^{\frac{1}{2}}|G/Z(G):F(G/Z(G))|^{-\frac{1}{2}}\leq |G/Z(G):F(G/Z(G))|^{-\frac{1}{2}}.$$

On the other hand, the bound of Wiegold [8, Chapter 4, p.126-127] gives

$$(*) \qquad d^{-\frac{1}{2}(1+log_2 d)}\leq |G'|^{-1},$$

then (i) is proved.

Lemma 2.7, combined with [4, Theorem 8 (i)] and Proposition 2.4, gives

$$|G'|^{-1}<cp(G)<cp(G/Z(G))<log_2(|G/Z(G):sol(G/Z(G))|)^{-\frac{1}{3}}.$$

As before we use $(*)$ and (ii) follows.

Lemma 2.7, combined with [4, Theorem 9] and Proposition 2.4, gives

$$|G'|^{-1} < cp(G) < cp(G/Z(G)) \leq |G/Z(G) : sol(G/Z(G))|^{-\frac{1}{2}}.$$

As before we use $(*)$ and (iii) follows.

Lemma 2.7, combined with [4, Theorem 12 (i)] and Proposition 2.4, gives

$$|G'|^{-1} < cp(G) < cp(G/Z(G)) \leq \frac{4l-7}{2^{l+1}}.$$

As before we use $(*)$ and (iv) follows.

Lemma 2.7, combined with [4, Theorem 12 (ii)] and Proposition 2.4, gives

$$|G'|^{-1} < cp(G) < cp(G/Z(G)) \leq \frac{p^l + p^{l-1} - 1}{p^{2l-1}}.$$

Now the bound [8, p.102] gives

$$(**) \qquad p^{-\frac{1}{2}m(m-1)} \leq |G'|^{-1},$$

then (v) follows. ◇

The conditions $(*)$ and $(**)$ in the proof of Theorem B are classical restrictions on $|G'|$ (see [9]) of an infinite group G. Recent developments can be found in literature: for instance, [9] improves $(*)$ using techniques of Combinatorial Group Theory (see [9, Theorems 1.1, 1.3, 1.4]). The same authors of [9] have continued to improve these bounds during the last twenty years.

Example 3.1. Let n be a positive integer and $G = E \times \mathbb{T}^n$, where $\mathbb{T}^n$ is the n-dimensional torus group and E is a finite non-abelian group. See [6, pp.11–17] and [6, Proposition 2.42] for details. Of course, G is a compact group and if we know $cp(E)$, then $cp(G) = cp(E)cp(\mathbb{T}^n) = cp(E)$ by an application of Lemma 2.1. This means that informations on $cp(G)$ can be deduced from those on $cp(E)$. Note that $cp(E)$ is well known by [1, 3, 4, 5, 7, 10], since E is a finite group. This construction gives a source of examples both for Theorem A and Theorem B. ◇

Acknowledgments

We thank the referee for some useful suggestions.

References

[1] P. Erdös and P. Turan, On some problems of statistical group theory, *Acta Math. Acad. Sci. Hung.* **19** (1968), 413–435.

[2] A. Erfanian and R. Kamyabi-Gol, On the mutually n-tuples in compact groups, *Int. J. Algebra* **(6)** Vol. 1 (2007), 251–262.

[3] P. X. Gallagher, The number of conjugacy classes in a finite group, *Math. Z.* **118** (1970), 175–179.

[4] R. M. Guralnick and G. R. Robinson, On the commuting probability in finite groups, *J. Algebra* **300** (2006), 509–528.

[5] W. H. Gustafson, What is the probability that two groups elements commute? , *Amer. Math. Monthly* **80** (1973), 1031–1034.

[6] K. H. Hofmann, and S. A. Morris, *The Structure of Compact Groups*, de Gruyter, Berlin, New York, 1998.

[7] M. S. Lucido and M. R. Pournaki, Elements with Square Roots in Finite Groups, *Algebra Colloq.* **12** (2005), 677–690.

[8] D. J. Robinson, *Finiteness conditions and generalized soluble groups*, vol. I and vol.II, Springer Verlag, Berlin, 1972.

[9] D. Segal and A. Shalev, On groups with bounded conjugacy classes, Quart. J. Math. Oxford **(2)**, 50 (1999), 505–516.

[10] G. J. Sherman, What is the probability an automorphism fixes a group element?, *Amer. Math. Monthly* **82** (1975), 261-264.

In: Mathematical Theories, Models and Groups ISBN 978-1-61324-250-6
Editor: Marcus J. Grecho

Chapter 3

ON HIGHER DERIVATIONS: A SURVEY

***Claus Haetinger*[1,*], *Mohammad Ashraf*[2,†] *and Shakir Ali*[3‡]**
[1]Center of Exact and Technological Sciences,
UNIVATES University Center, Lajeado-RS, Brazil
[2]Departament of Mathematics,
Aligarh Muslim University, Aligarh, India
[3]Departament of Mathematics,
Aligarh Muslim University, Aligarh, India

Abstract

We present an historical account on higher derivations, generalized higher derivations and other kinds of higher derivations in rings, based on the work of several authors. Moreover, recent results as well as some possible directions for future researches on these topics have been discussed in details. Finally, some examples and applications of higher derivations have been given.

1. Introduction

Many questions on derivations have been considered during the development of the ring theory, specially the relationships between derivations and the structure

*E-mail address: chaet@univates.br
†E-mail address: mashraf80@hotmail.com
‡E-mail address: shakir50@rediffmail.com

of rings, and they were already generalized in several directions (see the work of authors [4] for an historical account, examples and open problems). One of these directions reports to higher derivations. Rings with higher derivations are not the kind of subject that undergoes tremendous revolutions, however they have been studied in many papers during the last 60 years.

This paper is an attempt to present the higher derivations and its variants in such a light, and in a manner suitable for everybody who have some basic knowledge in ring theory. In order to make the treatment as self-contained as possible, and to bring together all the relevant material in a single paper, we have included several references. Much of the motivation for this paper is historical, and we have taken the opportunity to weave historical comments into the body of the text where it is appropriate.

Throughout, the present paper R will denote an associative ring with center $Z(R)$. For any $x,y \in R$, as usual $[x,y] = xy - yx$ and $x \circ y = xy + yx$ will denote the well-known Lie and Jordan products, respectively. A ring R is *n-torsion-free*, where n is an integer in case $nx = 0$, $x \in R$ implies $x = 0$. We shall make extensive use of basic commutator identities:

$$[xy,z] = x[y,z] + [x,z]y, \ [x,yz] = y[x,z] + [x,y]z$$

$$(x \circ (yz)) = (x \circ y)z - y[x,z] = y(x \circ z) + [x,y]z,$$

$$\text{and } ((xy) \circ z) = x(y \circ z) - [x,z]y = (x \circ z)y + x[y,z].$$

Recall that a ring R is *prime* if for any $a,b \in R$, $aRb = \{0\}$ implies that $a = 0$ or $b = 0$, and is called *semiprime* in case $aRa = \{0\}$ implies $a = 0$. An additive subgroup $U \subseteq R$ is said to be a *Lie ideal* (resp. *Jordan ideal*) of R if whenever $u \in U$ and $r \in R$, then $[u,r]$ (resp. $(u \circ r)$) is also in U. A Lie ideal such that $u^2 \in U$ for all $u \in U$ is called a *square closed* Lie ideal. Furthermore, if the Lie ideal U is square closed and $U \subseteq Z(R)$, then U is called an *admissible* Lie ideal. A map d: $R \to R$ is a *derivation* of a ring R if d is additive and satisfies the Leibniz' rule; $d(ab) = d(a)b + ad(b)$, for all $a,b \in R$. The most natural example of a nontrivial derivation is the usual differentiation on the ring $F[x]$ of polynomials defined over a field F. For a fixed $a \in R$, define $d : R \to R$ by $d(x) = [x,a]$ for all $x \in R$. Then, the function d so defined can be easily checked to be additive and $d(xy) = [xy,\ a] = x[y,\ a] + [x,\ a]y = xd(y) + d(x)y$, for all $x,y \in R$. Thus, d is a derivation which is called *inner derivation* of R associated with a and is usually denoted by I_a. An additive mapping $d : R \to R$ is called a *Jordan derivation* if $d(a^2) = d(a)a + ad(a)$ holds for all $a \in R$. It is obvious to see that every derivation on a ring R is a Jordan derivation but not conversly. An additive mapping $F : R \to R$ is called a *generalized derivation* (resp. *generalized Jordan derivation*) in case $F(xy) = F(x)y + xd(y)$ (resp. $F(x^2) = F(x)x + xd(x)$) holds for all $x, y \in R$ (resp. for all $x \in R$), where $d : R \to R$ is a derivation. An additive mapping $d : R \to R$ is said to be a *Jordan triple derivation* if $d(aba) = d(a)ba + ad(b)a + abd(a)$ holds for all $a, b \in R$. An additive mapping $f : R \to R$ is said to be a *generalized Jordan triple derivation* if there exists a Jordan triple derivation $d : R \to R$ satisfying $f(aba) = f(a)ba + ad(b)a + abd(a)$ for all $a, b \in R$. Clearly, every generalized derivation (resp. generalized triple derivation) on a ring is a generalized Jordan derivation (resp.generalized Jordan triple derivation). But the converse need not be true in general. A number of authors have studied this problem in the setting of prime and semiprime rings (viz. [3], [5], [6], [7], [13], [14], [22], [53], [54], [60], [62] and [72], where further references can be found). An historical development of the study of derivations in rings is given by authors in [4].

Let $\mathbb{N}$ be the set of natural numbers including 0 and $D = (d_i)_{i \in N}$ be a family of additive mappings of R such that $d_0 = id_R$. Then D is called a *higher derivation* (HD, for short) if for every $n \in \mathbb{N}$, we have

$$d_n(ab) = \sum_{i+j=n} d_i(a)d_j(b), \text{for all } a, b \in R \text{ and, } n \geq 1.$$

Also, D is called a *Jordan Higher Derivation*(JHD) if for every $n \in \mathbb{N}$, we have $d_n(a^2) = \sum_{i+j=n} d_i(a)d_j(a)$, for all $a \in R$; and D is called a *Jordan Triple Higher Derivation*(JTHD) if for every $n \in \mathbb{N}$, we have $d_n(aba) = \sum_{i+j+k=n} d_i(a)d_j(b)d_k(a)$, for all $a, b \in R$. Similarly, if U is a Lie ideal of R, then D is said to be a *HD (JHD, JTHD) of U into R* in case that the above corresponding conditions are satisfied for all $a, b \in U$.

1.1. Examples of Higher Derivations

In this section, we give some non-trivial examples of higher derivations.

Example 1.1.1. Let K be any field and define in $K[x]$, $d_i(x^j) = \binom{j}{i}x^{j-i}$, where $\binom{j}{i} = 0$ if $i > j$. Then $D = (d_i)_{i \geq 0}$ is a higher derivation on $K[x]$.

Example 1.1.2 ([41], Exemplo 1.4.2). Let $R = K[x]\langle y,z\rangle$ (K field, $char(K) \neq 2$) denotes the free ring with coefficients in $K[x]$ in the indeterminates y, z (where y and z commute with x but not with each other), a 2-torsion-free prime ring. Thus $Z(R) = K[x]$. Let $D = (d_n)_{n\in N}$ be a sequence of additive mappings of R such that $d_0 = id_R$ and, for every $n \geq 1$,

$$\begin{aligned} d_n(x) &= p_n(x), \text{ where } p_n(x) \in K[x]; \\ d_n(y) &= q_n(x,y,z), \text{ where } q_n(x,y,z) \in R; \\ d_n(z) &= r_n(x,y,z), \text{ where } r_n(x,y,z) \in R; \\ d_n(a^j) &= \sum_{i=0}^{n} d_i(a^{j-1})d_{n-i}(a), \text{for all } j \geq 2, \end{aligned}$$

where $a = x$, y, or z. Then D is a HD of R.

Reciprocally, every higher derivation of R is of this type for some $p_i(x)$, $q_i(x,y,z)$, $r_i(x,y,z)$.

This example can be easily generalized to rings with a finite number of indeterminates, that is, of the form $R = K[x_1,x_2,\ldots,x_s]\langle y_1,y_2,\ldots,y_m\rangle$.

Example 1.1.3 ([77]). Let R be a ring, and $a \in R$. If we put $d_0 = id_R$, and $d_n(x) = (-1)^n(xa^n - axa^{n-1}) = (-1)^n(xa - ax)a^{n-1}$ for $n \geq 1$, $x \in R$, then $D = (d_n)_{n\in N}$ is a higher derivation on R.

Some more examples can be found in [41]. Higher derivations of a ring R have been studied in many papers mainly in commutative rings, but also in noncommutative rings (see, for example, [12], [17], [22], [26], [27], [28], [31], [32], [33], [41], [42], [43], [44], [45], [47], [50], [64], [74], [76], [77], [82], [83], [84], [85], [89], [96], [98] and [100]).

Finally, let us say that the historical approach of this paper is partially based on work of several authors, while most of the sections are strongly based on the writers background and researches. Therefore, for many subjects treated here, we have only one superficial theoretical knowledge. We present some recent applications of higher derivations, as well as some problems that could be searched. We do not have the pretension to list here all the possible results and problems, neither to affirm that the cited results are the most importants. One more time, the interested readers can consult the innumerable references cited here.

2. Some Topics on Higher Derivations

2.1. Origin

The Higher Derivations (HD, for short) have been considered for the first time in 1936 by H. Hasse and F.K. Schmidt and independently by O. Teichmüller, being difficult to affirm with certainty which of them was the pioneer in this study. In what follows, we will see better why this seems to be truth.

The HD have its origin in the Differential Calculus in positive characteristic p, when Hasse [45] justified the theory of the HD in fields of algebraic functions. He used formulas to successive derivatives of implicit functions, known of the Analysis. In fields of characteristic zero it does not have difficulties, because the iterations of an usual derivation are used.

The paper of Hasse [45] is continuation of his another previous paper [44] of 1934, that deals with the HD in a field of algebraic functions K in just one indeterminate over a complete field of constants k, with characteristic p (where $p = 0$ or p is prime). In [45], the author generalizes results of [44].

O. Teichmüller [93] also avoided the complicated formulas of derivations of implicit functions in the field of positive characteristic. Without entering in details, Teichmüller viewed the fields of algebraic functions as inseparable extensions of invariant birational subfields. So, he could introduce the HD and its properties without using the development in power series. With this, a differential theory for the simplest inseparable extensions of fields of prime characteristic is obtained with a purely algebraic basement. In this direction, Teichmüller's paper was useful as a basement for works in p-algebras.

Now we present a stretch of a Schmidt's letter addressed to Hasse, that can be seen in [46].

> *"Vielleicht interessiert es Sie, daßich mir die von Ihnen unlängst entwickelte Theorie der höheren Differentialquotienten in einem algebraischen Funktionenkörper kurz vor Erscheinen Ihrer Arbeit*[1] *in etwas anderer form für meine gegenwärtige Vorlesung ebenfalls überlegt hatte. Mein Ziel war, diese Differentialquotienten sogleich im Großen einzuführen. Ich ging also nicht wie Sie von der Potenzrehenentwicklung an einer Stelle aus, weil ich gerne das Rechnen an der definierden Gleichung vermeiden wollte, das dann beim übergang vom Kleinen*

[1] This refers to the paper of Hasse [45].

zum Großen erfordelich ist. Aus der Schlußbemerkung Ihrer Abhandlung ersah ich später, daß auch O. Teichmüller eine Erklärung der höheren Differentialquotienten aufgestellt hat, die von vornherein im Großen verläuft, und ich nahm daher ohne weiters an, daß meine Definition mit der Teichmüller übereinstimme. Die soeben erschienene Arbeit Teichmüllers[2] *überzeugt mich aber, daß das doch nicht der Fall ist. Da meine Definition im Rahmen der Theorie der algebraischen Funktionen rasch zu den benötigten Ergebnissen führt, möchte ich sie Ihnen kurz mitteilen".* (25, Mai 1936)

Basically, Schmidt wrote to Hasse the following: *"Maybe warm you that I developed alone your theory of the differential quotients of higher order in a field of algebraic functions in a short while different way the one that your, for one future lecture. My main objective was introducing these differential quotients directly in an ampler sense. I did not use the development in power series in a point as you, because I would like to prevent the calculation of the defined equation, that would be necessary for the transition of the restricted one to the ample one. Of the conclusion of your work [45], I saw latter that the lecture of O. Teichmüller on the higher order differential theory is since the beginning in the ample direction and, therefore, I assumed that my definition was in accordance with him. The recent work of O. Teichmüller [93] convinced me, however, that this is not correct. As my definition of the theory of the algebraic functions it leads to the desired results, I would like briefly communicate you into it briefly what I did".*

Hasse presented a detailed approach of this theory [46], in which he disagreed with some points of his original argument. After concluding this article, Hasse knew that Schmidt had found an improvement and a generalization of the same one, and suggested the book "Zusatz bei der Korrektur" (pg., 223-237) to the readers for details.

With these remarks, we close the discussion concerning the origin of the Higher Derivations.

[2]This refers to the paper of Teichmüller [93].

2.2. Chronological Development

We know that the p-th power of a derivation in polynomial rings of prime characteristic p is zero. As an application of the classical theory of the differentials, based on Hasse, Schmidt and M. Deuring[3], in 1952 A. Jaeger [59] used these ideas to develop a modified theory of the algebraic differentials independently of the ring characteristic. So, it was possible to define non trivial higher derivations.

In 1966, R. Berger [11] published a paper showing what was developed until that referring date on the theory of the Higher Derivations, specially by Hasse and Schmidt. Among other results, he cited the work of Kähler, which proved that derivations, seen as one R-module, are very useful in several areas in Commutative Algebra and Algebraic Geometry. According to Berger, even so this approach does not bring no newness because the differential module of Kähler is the dual of the module of the derivations, the formularizations are in general significantly smoother in the language of the differential modules in this case. This construction was applied by Schmidt to inseparable field extensions. Following in this direction, Berger studied the universal derivations of a differential algebra. He also analyzed extensions K of fields k in characteristic p. Knowing that the first-order HD are related with the ones p-bases of K over k, he got that in HD of larger order we have p-bases of $K^{p^s}k$ over k, where $s = 1, 2, \ldots$. Furthermore, he calculated the inseparability exponent of these extensions, as Schmidt had calculated.

According to Hasse, Schmidt and Teichmüller, a Higher Derivation in a ring R is a sequence of additive applications $D = (d_i)_{i \in N}$ of R satisfying the conditions $d_0 = id_R$ and $d_n(ab) = \sum_{i=0}^{n} d_i(a)d_{n-i}(b)$ for all $a, b \in R$ and for all $n \in \mathbb{N}$.

As we wrote above, the concept of Higher Derivations was originally introduced in order to remove some of the anomalies in the calculus of derivations on fields of characteristic $p \neq 0$ (cf. [45], [59]). Nevertheless, HD have important applications in characteristic zero situations, as shown by N. Heerema [50] in a celebrated paper where he showed a review of the theory of HD on fields which form a background for the study of the uses of HD in automorphism theory of complete local rings. The abundance of HD in an algebra A is determined by the abundance of the derivations in A $\left(d \to d_n = \frac{d^n}{n!}\right)$. In 1927, R. Baer showed

[3]This refers to the book "Arithmetische theorie der algebraischen Funktionen".

that the subfields of a field h of characteristic zero which are fields of constants of derivations in h, are exactly the fields k that are algebraically closed in h [9]. Latter in 1960, Heerema [49] showed that the constant fields of Higher Derivations in h are the subfields of h which are algebraically closed in h. Moreover, he supposed that F is a perfect field. It is known [94] that in this case, F can be embedded in $F[[x]]$, its power series ring in the single indeterminate x, in only one way. Heerema proved that this unique embedding property can also be very easily obtained as a consequence of the fact that F is perfect only if it has no nontrivial derivations. If F has characteristic zero, the various ways in which F can be embedded in $F[[x]]$ can be determined constructively and biuniquely from Higher Derivations in F.

2.3. Kinds of Higher Derivations

There exist several other types of ring mappings that could be called Higher Derivations. Let us now see some of them, in order to better understand the chronological development and some of the applications of the HD.

2.3.1. *q*-th Order Derivations

In 1970, Y. Nakai [73] published a paper where he considered k and A as commutative rings with unity element, assuming that A is a k-algebra. Then he defined a *q-th order derivation* D of A into an A-module F over k as an element of $Hom_k(A, f)$ such that for any set of $(q+1)$-elements $(x_0, x_1, \ldots, x_q)$ of A we have the identity

$$D(x_0 x_1 \cdots x_q) = \sum_{s=1}^{q} (-1)^{s-1} \sum_{i_1 < \ldots < i_s} x_{i_1} \cdots x_{i_s} D(x_0 \cdots, \check{x}_{i_1} \cdots \check{x}_{i_s} \cdots x_q).$$

The first order derivation is just an ordinary derivation. This interesting notion of finite order higher derivations (FOHD, for short) seems to have been introduced by H. Osborn [78], in 1967, and by R.G. Heyneman and M.E. Sweedler [55], in 1969.

Nakai developed the theory of Kähler FOHD and extended them to the FOHD of C^∞ functions. In a subsequent paper, he applied this present theory to the Galois theory for purely inseparable field extensions of finite exponent[4].

[4] Paper published in Journal of Science of the Hiroshima University, series A-I, Vol 34 (1970), but we were unable to obtain its title.

The relatioship between this definition of FOHD and the HD definition of Hasse and Schmidt is the following: if $D = (d_i)_{i \in N}$ is a HD of rank finite (or infinite), then m-th component d_m is an m-th order derivation (in the sense of Osborn). But an m-th order derivation can not necessarily be an m-th component of a higher derivation. It would be an interesting problem to find a condition for an m-th order derivation (in the sense of Osborn) to be an m-th component of a higher derivation (in the sense of Hasse and Schmidt).

2.3.2. Locally Finite Iterative Higher Derivations

In 1978, Nakai [74] presented the *locally finite iterative higher derivations* (LFHD, for short) of a commutative ring A with unity element. If $D = (d_i)_{i \in N}$ is a Higher Derivation of A, then D is called locally finite if for any $a \in A$ there exists an index j such that $d_n(a) = 0$ for all $n > j$. Furthermore, D is called an *iterative HD* if $d_i d_j = \binom{i+j}{i} d_{i+j}$ for every $i, j \geq 0$. Nakai considered k an algebraically closed field and A an integral domain such that $A \supset k$ where $dim_k A = 2$. Then he obtained a characterization of two-dimensional polynomial rings. As a corollary, a characterization of a line in an affine plane is meant a curve C which can be taken as a coordinate axis of A^2. He called a curve C; $f(x,y) = 0$ a *quasi-line* if the coordinate ring $k[x,y]/\langle f \rangle$ is isomorphic to one-parameter polynomial ring. It is known that of the ground field is the complex number field $\mathbb{C}$, then a quasi-line is always a line (cf. [1]). Combined with Nakai results it turns out that if the plane curve C: $f(x,y) = 0$ is a quasi-line over $\mathbb{C}$, then the derivation $d_f = (\partial f / \partial y)\frac{\partial}{\partial x} - (\partial f / \partial x)\frac{\partial}{\partial y}$ is locally nilpotent, i.e., the higher derivation $\left(1, d_f, \frac{1}{2!}d_f^2, \ldots\right)$ is a LFHD and vice-versa. The direct proof of this fact was expected very much in Algebraic Geometry.

2.3.3. Commuting Finite Schmidt Higher Derivations

It is well known that in certain problems involving fields of characteristic $q \neq 0$, Higher Derivations, i.e., derivations of infinite rank, often yield results which can not be obtained for ordinary derivations. So it is natural to study HD and their relationships to ideals in commutative rings.

In 1970's, K.R. Goodearl published several papers and derived formulas for the global homological dimension of the ring of formal linear differential operator over a commutative noetherian ring with a finite set of commuting derivations ([37], [38], [39]). Later, in the year 1983, formulas for the Krull dimension of the ring of formal linear differential operators over a commutative noetherian ring with a finite set of commuting derivations were also derived by Goodearl and T.H. Lenagan [40]. Further in the year 1989, D.R. Malm [68] derived formulas for the global homological dimension of the ring of differential operators over a commutative noetherian ring with a commuting finite Schmidt higher derivation[5]. Also, he derived formulas for the Krull dimension (in the sense of Gabriel and Rentschler) of the ring of differential operators over a commutative noetherian ring R with a commuting finite Schmidt higher derivation when R is a$\mathcal{Q}$-algebra or when all residue fields of R have positive characteristic. One year earlier, he developed necessary and sufficient conditions for certain differential operator rings to be simple. He did this for a ring with finitely many commuting derivations and for a ring with a commuting Schmidt higher derivation of finite length. Malm also obtained a correspondence between finite sets of derivations and Schmidt higher derivations [69]. These Malm's results have close connection with algebraic derivation problems and problems related with HD satisfying linear relations (see [26], [33], [67], [79], [95]).

2.3.4. Higher Derivations as Integrals of Derivations

We want to introduce one more motivation in the study of HD, based on D. Farkas, C. Geiss and E.N. Marcos [31].

In 1982, H. Matsumura [70] defined the integral derivations in commutative rings. If A is a commutative ring with unity element and d is an usual derivation of A, we say that d is an *integral* derivation if there exists an homomorphism $E : A \to A[[t]]$ such that $E(a) \equiv a + td(a) \pmod{t^2}$. Equivalently, d is said to be integral if there exists a HD $D = (d_i)_{i \in N}$ such that $d_1 = d$. In this case, D is said to be one *integral* of d. Integral derivations have many good properties. Roughly speaking, many undesirable derivation phenomena in characteristic p disappear if we consider only integrable derivations.

A. Seidenberg ([86], [87], [88]) showed necessary conditions for integrability of derivation. Y. Nakai ([73], [74]) proved enough conditions. Already Matsumura considered smooth algebras- or strongly smooth algebras -and did use of the theory of Andre homology [2]. Moreover, he showed that there exist many integrable derivations (in case where A is a finitely generated integral domain over a perfect field). If $A \supset \mathcal{Q}$, then every derivation is integrable $\left(d_n = \frac{d^n}{n!}\right)$.

[5]D.R. Malm mad use of Schmidt higher derivations with the same meaning as we use Higher Derivation.

The same occurs if A is a field of arbitrary characteristic ([70], Theorem 6). But, in general, there are nonintegrable derivations.

The study of the integrable derivations also applies to noncommutative rings. The definition of integrable derivation can be given by one of the following three equivalent conditions, whose proofs can be easily obtained in the literature: (a). there exists an homomorphisms of k-algebras $f : A \to A[[t]]$ such that $f(a) \equiv a + td(a) \pmod{t^2}$; (b). there exists a $k[[t]]$-automorphism $g : A[[t]] \to A[[t]]$ such that for every $a \in A$ we have $g(a) \equiv a + td(a) \pmod{t^2}$; (c). there exists a HD $D = (d_i)_{i \in N}$ in A such that $d_1 = d$.

In this direction, we detach the paper of Farkas, Geiss and Marcos [31]. The most naive way to understand a finite-dimensional associative algebra is to find a base and to analyze its multiplication table. Now a days, we consider the scheme of all associative n-dimensional algebras over a field k as a subscheme of the affine space $Hom_k(k^n \otimes k^n, k^n)$. Then $GL_n(k)$ acts on the k-rational scheme points in such a way that the orbits can be seen as isomorphisms classes of n-dimensional algebras. The stabilizator of a point can be identified with the automorphism group scheme of the corresponding algebra. The geometry in the point is particularly good when the scheme is smooth. The automorphism group scheme is automatically smooth when the characterisitc of k is zero, since the classic characterization of commutative Hopf algebras. If k has positive characteristic, then this is a little more mysterious. The main contribution of the authors was to give a simple and practical reformularization of the smooth algebras. They had proved that the automorphism group scheme of a finite-dimensional algebra A over a perfect field k is smooth if, and only if, each k-derivation of A is integrable ([31], Theorem 1.2). This criterion gave a purely algebraic characterization for an Algebraic Geometry fact. More details can be found in ([29], [30], [34], [70], [91]).

2.3.5. Generalized Higher Derivations

The notion of generalized derivations on a ring A which was introduced by M. Brešar [15] is related to a derivation of A. In [71], A. Nakajima defined another type of generalized derivations without using derivations, and gave some categorical properties of that generalized derivations. When A has an identity element, these two notions coincide. The results in [71] were extended to generalized Jordan and Lie derivations in [72].

On the other hand, in his paper [84], P. Ribenboim gave some properties of higher derivations of modules. His higher derivation f from an A-module M to M is defined by using a Higher Derivation $D = (d_t)$: $A \to A$ and the case of length 1 is nothing but a generalized derivation in the sense of Brešar whenever d_0 is the identity map on A.

In 2000, A. Nakajima defined a generalized higher derivation ([72], Definition 2.3) without using a HD at the viewpoint of [71] and gave some categorical properties which are related in [71]. He also treated generalized higher Jordan and Lie derivations. Let $F = (f_i)_{i\in N}$ be a family of additive mappings of R such that $f_0 = id_R$. Then, following W. Cortes and C. Haetinger [22], F is called: (i) a *generalized higher derivation* (GHD, for short) if there exists a HD $D = (d_i)_{i\in N}$ of R such that for every $n \in N$, we have $f_n(ab) = \sum_{i+j=n} f_i(a)d_j(b)$, for all $a, b \in R$; (ii) a *generalized Jordan higher derivation* (GJHD, for short) if there exists a HD $D = (d_i)_{i\in N}$ of R such that for every $n \in N$, we have $f_n(a^2) = \sum_{i+j=n} f_i(a)d_j(a)$, for all $a, b \in R$; (iii) a *generalized Jordan triple higher derivation* (GJTHD, for short) if there exists a HD $D = (d_i)_{i\in N}$ of R such that for every $n \in N$, we have $f_n(aba) = \sum_{i+j+k=n} f_i(a)d_j(b)d_k(a)$, for all $a, b \in R$. Similarly, if U is a Lie ideal of R, then F is said to be a *GHD (GJHD, GJTHD) of U into R* in case that the above corresponding conditions are satisfied for all $a, b \in U$.

Clearly every generalized higher derivation on R is a generalized Jordan higher derivation. But the converse statement does not hold in general. It is shown in [22] that if R is a 2-torsion-free ring which has a commutator right nonzero divisor and U is a square closed Lie ideal of R, then every generalized Jordan higher derivation on U into R is a generalized higher derivation. In a recent paper, Y.-S. Jung [62] improve Jing and Lu's result ([60], Theorem 3.5) for generalized Jordan triple derivations to generalized Jordan triple higher derivations, proving that if R is a 2-torsion-free prime ring, then every generalized Jordan triple higher derivation on R is a generalized higher derivation.

To conclude this section, let us mention few open problems concerning to GJHD and GJTHD:

Problem 2.3.5.1. *Let R be a 2-torsion-free ring and U be an arbitrary Lie ideal of R. Then every GJTHD of R (resp. of U into R) is a GHD of R (resp. of U into R).*

Problem 2.3.5.2. *Let R be a 2-torsion-free semiprime ring (resp. prime ring* *of $char(R) \neq 2$ and U an arbitrary Lie ideal of R. Then every GJTHD of R (resp. of U into R) is a GHD of R (resp. of U into R).*

Problem 2.3.5.3. *Let R be a 2-torsion-free ring and J be a Jordan ideal of R. Then every GJTHD of R (resp. of J into R) is a GHD of R (resp. of J into R).*

A more precise description of the generalized higher derivations, generalized Jordan higher derivations and generalized Jordan triple higher derivations need a lot of space, so we feel that it is better to resist the tempation to express this subject in greater details and instead refer to [22], [27], [28], [72] and to some of the most recent article [32], [33], [62] for the advanced theory.

2.3.6. Higher (U,R)-Derivations

Let U be a Lie ideal of a ring R. Following A.K. Faraj, C. Haetinger and A.H. Majeed ([27], [28]), an additive mapping $d: R \to R$ is said to be an (U,R)-*derivation* of R if $d(ur+su) = d(u)r + ud(r) + d(s)u + sd(u)$, for all $u \in U$, for every $r,s \in R$. Let F be an additive mapping of R into itself. We call F a *generalized (U,R)-derivation* of R if there exists an (U,R)-derivation d of R such that $F(ur+su) = F(u)r + ud(r) + F(s)u + sd(u)$, for all $u \in U$, for every $r,s \in R$. Let $D = (d_i)_{i\in N}$ be a family of additive mappings of R into itself such that $d_0 = id_R$. Then, D is said to be a *higher (U,R)-derivation* of R if for every $n \in \mathbb{N}$ we have $d_n(ur+su) = \sum_{i+j=n} d_i(u)d_j(r) + d_i(s)d_j(u)$, for all $u \in U$, $r,s \in R$. Further, let $F = (f_i)_{i\in N}$ be a family of additive mappings of R into itself such that $d_0 = id_R$. Then, F is called a *generalized higher (U,R)-derivation* if there exists a higher (U,R)-derivation $D = (d_i)_{i\in N}$ of R such that, for all $n \in \mathbb{N}$, $f_n(ur+su) = \sum_{i+j=n} f_i(u)d_j(r) + f_i(s)d_j(u)$ holds for all $u \in U$ $r,s \in R$.

Example ([28]): Let $R = \mathcal{M}_2(S)$ over a commutative ring S, $char(S) = 2$.
Let $U = \left\{ \begin{pmatrix} x & y \\ y & x \end{pmatrix} : x,y \in S \right\}$. Then U is a Lie ideal of R.
For all $n \in \mathbb{N}$, define d_n: $R \to R$ by

$$d_n\left(\begin{pmatrix} a & b \\ c & d \end{pmatrix}\right) = \begin{cases} \begin{pmatrix} 0 & -b \\ nc & 0 \end{pmatrix} & , if\ n = 1,2 \\ 0 & , if\ n \geq 3 \end{cases} \text{, for every } \begin{pmatrix} a & b \\ c & d \end{pmatrix} \in R.$$

Then D is a higher (U,R)-derivation.

Let $F = (f_i)_{i\in N}$ be a family of mappings of R into R defined by

$$f_n\left(\left(\begin{array}{cc} a & b \\ c & d \end{array}\right)\right)=\begin{cases}\left(\begin{array}{cc} na & 0 \\ 0 & -d \end{array}\right) & , if\ n=1,2 \\ 0 & , if\ n\geq 3\end{cases}, \text{ for every } \left(\begin{array}{cc} a & b \\ c & d \end{array}\right)\in R.$$

Therefore, using the higher (U,R)-derivation $D=(d_i)_{i\in N}$ which was defined earlier, F is a generalized higher (U,R)- derivation.

Very recently, ([28], Theorem 5.1) Faraj, Haetinger and Majeed considered generalized higher (U,R)- derivation of R and obtained the following result: Let R be a prime ring of char$R\neq 2$, U be an admissible Lie ideal of R. If $F=(f_i)_{i\in N}$ is a generalized higher (U,R)-derivation of R, then $f_n(ur)=\sum\limits_{i+j=n} f_i(u)f_j(r)$, for all $u\in U,\ r\in R,\ n\in I\!N$. This result extends Awtar's main theorem ([7], Theorem).

Here is a related open problem that awaits solution:

Problem 2.3.6.1. *Let R be a 2-torsion-free semiprime ring (resp. prime ring of char $R\neq 2$ and U an arbitrary Lie ideal of R). Then every generalized Jordan triple (U,R)-higher derivation of R (resp. of U into R) is a generalized (U,R)-higher derivation of R (resp. of U into R).*

2.4. Some Applications and Some Problems

The theory of higher derivations leading to applications in the automorphism theory of complete local rings. The theory of higher derivations does not play an important role only in ring theory, but also in Galois theory of fields (viz. [23], [24], [49], [51], [98]) as well as the theory of universal higher derivations and in separability criteria ([11], [80]). An extensive and deep theory has been developed especially for higher derivations of Lie algebras over field of characteristic zero ([58], Chapter 9), and Galois Theory for a class of purely inseparable field extensions (see; e.g., [23], [24], [92]). To explain more precisely, we have the following:

2.4.1. Higher Derivations Applied to Automorphisms of Complete Local Rings

The higher derivations in fields leads to several uses of HD in the theory of automorphisms of complete local rings. In fact, it is known that the HD can be applied to the Galois Theory of fields ([23], [24], [92], [98]), to the theory of the universal HD [11], and to the inseparability criteria ([11], [80]).

Now, let A be an (nonassociative) algebra over a field k of zero characteristic, and d: $A \to A$ be a derivation. We remember that the set of all derivations of A ($Der(A)$) forms an additive group. If A is complete in some topology in which $D = (d_i)_{i \in N}$, where $d_i = \frac{d^i}{i!}$ for every $i > 0$, converges, then the exponential d, $exp(d) = \sum_{i=0}^{\infty} \frac{d^i}{i!}$, is an automorphism of A. For example, if $k = \mathbb{R}$ (the set of real numbers) and $dim_{\mathbb{R}} A < \infty$, then $\mathbb{R}$ is complete in the Euclidean topology and $exp(d)$ is a convergent series. Therefore, $exp(d)$ is an automorphism of A, for each $d \in Der(A)$. In general, if d is nilpotent and associative, then $exp(d)$ contains only a finite number of nonzero elements, so it is an automorphism, too. These automorphisms have an important role in the structure of the theory of Lie Algebras over zero characteristic fields ([58], Chapter 9). As well as in the case of derivations, it is known that all HD defined in an integral domain has only one extension to the quotient field. N. Heerema generalized this property, constructing a HD in a complete regular local ring with residual field in characteristic not zero ([47], Theorem 16, [50]). It is possible to obtain a HD group isomorphic to an inertial automorphisms group of a complete local ring. R.L. Davis obtained a generalization of the Theory of Galois in purely inseparable extensions of exponent 1 of fields for purely inseparable extensions of arbitrary finite exponent ([23], [24], [56]).

Higher Derivations are also useful for the aproximation of inertial automorphisms with automorphisms of derivations in M-adic topologies (where M is the maximal ideal in a complete local ring R). More applications in this kind of rings can be seen in ([23], [48], [50], [51]). In ([50]), N. Heerema showed methods of constructing automorphisms using HD, particularly in connection with the problem of identifying the factor groups of the higher ramification series of a complete local ring.

An interesting application of the use of Higher Derivations in commutative rings was obtained in the year 1994 by P. Brumatti and Y. Lequain [19]. For a complete local ring (R, M) (where R is a noetherian commutative ring with unity with M as unique maximal ideal) containing the field of rational numbers $\mathbb{Q}$, a derivation d of R such that $d(M) \not\subseteq M$, and for the biggest d-ideal $I_M(d)$ contained in M, M. Zariski ([99], Lemma 4) has shown that: (a). there exists a subring R_0 and an element $u \in M$ analytically independent over R_0 such that $R = R_0[[u]]$; (b). the ring $R/I_M(d)$ is isomorphic to the ring of formal power series $(R/M)[[X]]$. In particular, $I_M(d)$ is a prime ideal of dimension 1. In characterisitc $p \neq 0$, these results have no chance to be true and, instead of a derivation, the natural concept to consider is that of an iterative differentiation (Higher Derivation) $D = \{d_i\}_{i\geq 0}$. In this setting, B. Singh [89] has generalized Part (a) of Zariski's result. Brumatti and Lequain generalized Part (b) in commutative rings with unity, showing that $R/I_M(D)$ is also isomorphic to the ring of formal power series $(R/M)[[X]]$. However, Zariski's method of proof could not be adapted to the new setting, and they had to develop a new approach that uses deep results of Chavalley and Nagata on complete local rings. Besides, Brumatti and Lequain considered a local ring (R,M) with completion $(\hat{R},\hat{M})$ together with an iterative differentiation (Higher Derivation) $D = \{d_i\}_{i\geq 0}$ and they studied the filtration defined by $M_D^{(j)} = \{x \in R \mid d_i(x) \in M, \forall i \leq j-1\}$.

In 1972, W. Brown and W. Kuan [17] had also found a generalization of Zariski's result in this sense, which is quite used in finitely generated commutative rings, that is, those rings which naturally arise in Algebraic Geometry. The main result ([17], Theorem 6) states that if a HD, $D = (d_i)_{i\in N}$, is such that d_1 is invertible in R and $d_i(u) = 0$ for every $i > 1$ with $u \in M$ (fixed), then there exists a subring R_0 of R such that R_0 is a complete local ring, u is analytically independet over R_0, and $R = R_0[[u]]$.

Also related with local rings, P. Brumatti, A. Doering and Y. Lequain [18] had made the following study: for a field k and a local ring (R,M) that is a localization of a finite k-algebra, A. Seidenberg [88] showed that if the characteristic of k is zero and R is D-simple for a subset D of k-derivations of R, then R is regular. It had been conjectured that for any local ring, differential simplicity should imply regularity. However, examples of non-regular local rings that contain the rational numbers$\mathbb{Q}$ and are d-simple for a certain derivation d have been constructed in ([25], [40] and [66]). Also, in characteristic $p \neq 0$, examples of non-regular local rings that are D-simple for a certain Hasse-Schmidt derivation (Higher Derivation) D have been constructed in [16]. All these examples have dimension 1, or are polynomial rings over rings of dimension 1; in particular, all of them are Cohen-Macaulay. Brumatti, Doering and Lequain constructed a 2-dimensional local domain of arbitrary characteristic that contains a field and that is D-simple for a certain Hasse-Schmidt derivation D, but that has grade 1, hence that is not Cohen-Macaulay. Among other results, for a local ring (R,M) of arbitrary characteristic that is D-simple for a set D of Hasse-Schmidt derivations, they studied the connection between the grade of R and the set of prime divisors of (0) in the completion $\hat{R}$. Furthermore, they proved that the henselizations[6] of a Hasse-Schmidt differentially simple local ring is also Hasse-Schmidt differentially simple. Then, they obtained that the henselizations of the local domains described in the paper are henselian local domains that are Hasse-Schmidt differentially simple, not Cohen-Macaulay and such that in their completion (0) has an embedded prime divisor of coheight 1. This shows that, for arbitrary characteristic, Hasse-Schmidt differentially simple rings can have a pretty bad behaviour with respect to regularity, even if they are henselian.

2.4.2. Herstein's Problem for Higher Derivations

In 1950's, I. N. Herstein initiated the study of the relationship between the associative, Jordan and Lie structure of associative rings. We refer the reader to ([4], [52], [53], [54], [78], [31]), where one can find further references and more detailed explanations concerning the motivation and the background of these researches.

One can verify that a Jordan derivation in an associative ring R is a derivation on the Jordan ring under the induced Jordan multiplication. Note that the definition of Jordan derivation presented in the work of Herstein is not as the given above. In fact, Herstein constructed, starting from the ring R, a new ring, namelly the Jordan's ring of R, defining the product in this one as being $a \circ b = ab + ba$, for any $a, b \in R$. Clearly, this new product is well-defined and it can be easily verified that $(R, +, \circ)$ is a ring. So, an additive mapping d, from the Jordan's ring into itself, is said by Herstein to be a Jordan derivation, if $d(a \circ b) = d(a) \circ b + a \circ d(b)$, for every $a, b \in R$. However, in the year 1957, Herstein proved a classical result in this direction which becomes a jumping point for many workers latter. The result to which we refere is namely:

Theorem 2.4.2.1. ([53], Theorem 3.1) *If R is a prime ring of characteristic different from 2, then every Jordan derivation of R is a derivation.*

Based on M. Brešar [13] and on R. Awtar [7], M. Ferrero and C. Haetinger [32] extended Herstein's theorem for HD. In the main result ([32], Theorem 1.2) the authors proved the following: Let R be a 2-torsion-free semiprime ring (resp. prime ring of characteristic different of 2 and U an admissible Lie ideal

[6]A field K with a valorization v is said to be *henselian* if every algebraic extension of K allows exatcly one extension of v. It is shown that this property is equivalent to the Hensel's Lemma in (K, V). The smallest extension of a valorizated field (K, v) which is a henselian field is said a *henselization* of K. It is shown that this extension is unique.

of R). Then every JTHD of R (resp. of U into R) is a HD of R (resp. of U into R). Moreover, ([32], Theorem 1.3) states that: Let R be a 2-torsion-free ring and U a Lie ideal of R. Then every JHD of R (resp. of U into R) is a JTHD of R (resp. of U into R). The following is clear ([32], Corollary 1.4): Let R be a 2-torsion-free semiprime ring (resp. prime ring of characteristic different of 2 and U an admissible Lie ideal of R). Then every JHD of R (resp. of U into R) is a HD of R (resp. of U into R). In the same year, C. Haetinger [43] published a brief proof of this last result, without using JTHD. Further in 2003, M. Ashraf and N. Rehman [5] showed that in a 2-torsion-free ring R which has a commutator that is not a divisor of zero, every generalized Jordan derivation on R is a generalized derivation, and in 2005, W. Cortes and C. Haetinger [22] improved this result to a generalized Jordan higher derivation, that is, if R is a 2-torsion-free ring which has a commutator that is not a right divisor of zero and U is a square closed Lie ideal of R, then every generalized Jordan higher derivation of U into R is a generalized higher derivation of U into R.

The following problems are open till date:

Problem 2.4.2.1. *Let R be a 2-torsion-free semiprime ring (resp.prime ring of char $R \neq 2$ and U an arbitrary Lie ideal of R). Then every JTHD of R (resp. of U into R) is a HD of R (resp. of U into R).*

Problem 2.4.2.2. *Let R be a 2-torsion-free and U an arbitrary Lie ideal of R. Then every JHD of R (resp. of U into R) is a HD of R (resp. of U into R).*

2.4.3. Higher Derivations Satisfying Linear Relations

Recall that if $d : R \to R$ is a derivation of an algebra R over the field of rational numbers$\mathbb{Q}$, then putting $d_i = \frac{d^i}{i!}$ we have that $D = (d_i)_{i \in N}$ is a HD of R. Thus the derivation d is algebraic over R if and only if there exists a relation $\sum_{i=0}^{n} r_i d_i(x) = 0$, for all $x \in R$, and some elements $r_0, r_1, \ldots, r_n \neq 0$ in R.

In 1984, A. Nowicki [77] defined inner derivations of higher order of a ring R and proved that they correspond to the inner automorphisms of a suitable ring. Moreover, he proved that any HD of R is inner if an only if any usual derivation of R is inner. For Higher Derivations, the algebraicity situation send us to the study of HDs satisfying linear relations. Assume that R is an algebra of finite dimension over a field K. Then every HD, $D = (d_i)_{i \in N}$, where d_i is K-linear for every i, satisfies a K-linear relation and then an R-linear relation, since $End_K(R)$ is of finite dimension over K. Consequently, it is natural to study Higher Derivations which satisfy linear relations on a semiprime ring R.

This problem was studied by M. Ferrero and C. Haetinger ([33], [41]) in 2000-2002. Assume that R is a semiprime ring and $D = (d_i)_{i \in N}$ is a HD of R, Q is the left Martindale's quotient ring of R, and C is the extended cenroid of R (the center of Q). Then there is a unique extension $D^\star = (d_i^\star)_{i \in N}$ of D to a HD of Q ([41], Teorema 1.5.1). One of the main results of [33] states that the following conditions are equivalent: there exists an R-linear (Q-linear) relation for D on R; there exists an R-linear (Q-linear) relation for $D^\star$ on Q. This is an extension of the result mentioned in the algebraic derivation subsubsection above. The corresponding relation with C-algebraicity is no more true in this case. Also, if R is a prime ring and $D = (d_i)_{i \in N}$ is a HD of R which satisfies a minimal linear relation of the type $\sum_{i=0}^{n} r_i d_i(x) = 0$, for all $x \in R$, where $r_i \in R$ and $r_n \neq 0$, the authors gave a description of $d_1, d_2, \ldots, d_n$ in terms of X-inner derivations of R. This result is an extension of a V.K. Kharchenko' s result [63] which states that an algebraic derivation is X-inner, provided that $char(R) = 0$ or $char(R) = p$ is greater than the degree of algebraicity of d.

2.4.4. Higher Derivations in Incidence Algebras

Another interesting application is the study of the relatioship among the automorphisms and the Higher Derivations of incidence algebras. As we wrote above, the study of automorphisms and derivations of algebras has a long history. Several authors have contributed, for example, to the following three topics:

- automorphisms and derivations of incidence algebras of locally finite partially ordered or pre-ordered sets ([8], [75],[97]);
- automorphisms and derivations of upper triangular matrix algebras ([20], [61], [65]) and of full matrix algebras ([10], [21], [90]);
- Higher Derivations on (such) algebras ([57], [76], [77]).

M. Koppinen [64] studied the structure of these applications and actions. He considered first, a little more general situation, algebra homomorphisms between two incidence algebras, and then he deduced some corollaries for the above situations. This yield extensions of several older results of a similar type.

2.4.5. Higher Derivations of Modules

In 1980, P. Ribenboim [84] extended to modules his theory of higher order derivations of rings, introduced in ([82], [83]). Higher order derivations are families of maps between modules such that, among other properties, their zero components are module-homomorphisms. Therefore, they are bound to provide a finer information of "infinitesimal nature" on modules, than what may be achieved with the exclusive use of homomorphisms.

In [84], P. Ribenboim developed only the basic tools of the theory, creating a framework for further investigation. He introduced concepts like the group of invertible derivations of a module, which is a counterpart of the group of automorphisms. It serves to construct the graded Lie rings of derivations, canonically attached to any module. The existence of universal derivations and of the module of differentials allows to linearize the questions related to derivations. He defined also inner derivations of higher order, the exponential of a derivation and proceeded to study the kernel, image and exact sequences.

Finally, he considered derivation-smooth modules and derivation-projective modules, illustrating the contention that derivations do provide a finer method for analysing modules. Indeed, he was able to define the class of d-projective modules, which contains that of free modules and is contained in the class of projective modules.

Let Mod_R will denote the category of unitary right R-modules and τ be an hereditary torsion theory on Mod_R. S.H. Rim has shown in [85] that under certain conditions a higher anti-derivation d: $M \to M$ can be extended to a higher anti-derivation d^{τ}: $Q_{\tau}(M) \to Q_{\tau}(M)$, where Q_{τ}: $Mod_R \to Mod_R$ is the localization functor [36] at an hereditary torsion theory τ on Mod_R. By selecting the involution on the ring in the definition of a higher anti-derivation to be the identity mapping on R, a higher anti-derivation d: $M \to M$ becomes a higher derivation as defined by Ribenboim in [84]. Thus Rim's results, which generalize the results of J.S. Golan [35], showed that a higher derivation d: $M \to M$ can be extended to a higher derivation d^{τ}: $Q_{\tau}(M) \to Q_{\tau}(M)$ whenever the conditions of his proposition are met. Uniqueness of extensions of higher (anti-)derivations and the necessary and sufficient conditions for the existence of these extensions were not addressed. In [12], P.E. Bland introduced higher differential torsion theories and proved that a higher derivation d: $M \to M$ can always be extended uniquely to $Q_{\tau}(M)$, the module of quotients of M if and only if τ is a higher differential torsion theory on Mod_R. In the same paper, he also proved that a higher derivation d: $M \to M$ can be lifted uniquely to the module of coquotients $C_\tau(M)$ of M at a TTF theory on Mod_R. For more related results we refer to reader, ([35], [78], [84], [85]), and references therein.

References

[1] S. Abhyankar, T. Moh, Embedding of the line in the plane, *J. Reine Angew. Math.* **276** (1975), 148-166.

[2] M. André, *Homologie des algèbres commutatives,* Springer, (1974).

[3] M. Ashraf, A. Ali, S. Ali, On Lie ideals and generalized (θ, φ)-derivations in prime rings, *Comm. Algebra* **32** (8) (2004), 2977-2985.

[4] M. Ashraf, S. Ali, C. Haetinger, On derivations in rings and their applications, *The Aligarh Bull. Math.* **25** (2) (2006), 79-107.

[5] M. Ashraf, N. Rehman, On Jordan generalized derivations in rings, *Math. J. Okayama* **42** (2000), 7-9.

[6] M. Ashraf, N. Rehman, S. Ali, On Lie ideals and Jordan generalized derivations of prime rings, *Indian J. Pure & Appl. Math.* Vol. 34(2) (2002), 291-294.

[7] R. Awtar, Lie ideals and Jordan derivations of prime rings, *Proc. Amer. Math. Soc.* **41**(1984), 67-74.

[8] K. Baclawski, Automorphisms and derivations of incidence algebras, *Proc. Amer. Math. Soc.* **36** (1972), 63-74.

[9] R. Baer, Algebraische Tehorie der differentierbaren Funktionenkörper, S.-B. Heidelberger *Akad. Wiss. Abh.* **8** (1927), 15-32.

[10] G.M. Benkart, J.M. Osborn, Derivations and automorphisms of nonassociative matrix algebras, *Trans. Amer. Math. Soc.* **263** (1981), 411-430.

[11] R. Berger, Differentiale Höherer Ordnung und Körpererweiterungen bei Primzahlcharakteristik, S.-B. Heidelberger *Akad. Wiss. Math.-Natur. Kl.* (1966), 143-202.

[12] P.E. Bland, Higher derivations on rings and modules, *Int. J. Math. and Math. Sci.* **15**(2005), 2373-2387.

[13] M. Brešar, Jordan derivations on semiprime rings, *Proceedings of the AMS* **104** (4) (1988), 1003-1006.

[14] M. Brešar, Jordan mappings of semiprime rings, *J. Algebra* **127** (1989), 218-228.

[15] M. Brešar, On the distance of the composition of two derivations to the generalized derivations, *Glasgow Math. J.* **33** (1991), 89-93.

[16] W. Brown, Differantially simple rings, *J. Algebra* **53** (1978), 362-381.

[17] W. Brown, W. Kuan, Ideals and higher derivations in commutative rings, *Canad. J. Math.*, Vol. XXIV, No. 3 (1972), 400-415.

[18] P. Brumatti, A.M. de S. Doering, Y. Lequain, Differential simplicity, Cohen-Macaulayness and formal prime divisors, *Journal of Pure and Applied Algebra* **81** (1993), 1-11.

[19] P. Brumatti, Y. Lequain, Maximally differential ideals, *J. Algebra* **168** (1) (1994), 172-181.

[20] S.P. Coelho, Automorphism groups of certain algebras of triangular matrices, *Arch. Math.* **61** (1993), 119-123.

[21] S.P. Coelho, C.P. Milies, Derivations of upper triangular matrix rings, *Linear Algebra Appl.* **187** (1993), 263-267.

[22] W. Cortes, C. Haetinger, On Jordan generalized higher derivations in rings, *Turkish Journal of Mathematics* **29** (1) (2005), 1-10.

[23] R.L. Davis, *A Galois theory for a class of purely inseparable fields extensions,* Dissertation, Florida State University, Talahassee, Fla., 1969.

[24] R.L. Davis, A Galois theory for a class of purely inseparable extensions exponent two field extensions, *Bull. Amer. Math. Soc.* **75** (1969), 1001-1004.

[25] A.M. de S. Doering, Y. Lequain, Maximally differential prime ideals, *J. Algebra* **101** (1986), 403-417.

[26] F. Dumas, R. Vidal, Dérivations, et hautes dérivations dans certains corps gauches de Séries de Laurent, *Pacific J. Math.* **153** (2) (1992), 277-288.

[27] A.K. Faraj, *On higher derivations and higher homomorphisms of prime rings*, Ph. D. Thesis, Bagdad University (2006).

[28] A.K. Faraj, A.H. Majeed, C. Haetinger, *Generalized higher* (U,R)-*derivations,* (Submitted)

[29] D.R. Farkas, C. Geiss, E.N. Marcos, Diagonalizable derivations of finite-dimensional algebras I, *Israel Journal of Mathematics* **117** (2000), 157-181.

[30] D.R. Farkas, C. Geiss, E.N. Marcos, Diagonalizable derivations of finite-dimensional algebras II, *Pacific Journal of Mathematics* **196** (2) (2000), 341-351.

[31] D.R. Farkas, C. Geiss, E.N. Marcos, Smooth automorphism group schemes, In *Proc. on Representations of Algebras*, *International Conference on Representation of Algebras,* São Paulo (Brazil), July 19-24, (1999); Marcel Dekker, Inc.: New York, *Lectures Notes in Pure and Applied Math.* **224** (2002).

[32] M. Ferrero, C. Haetinger, Higher derivations and a theorem by Herstein, *Quaestiones Mathematicae* **25** (2) (2002), 249-257.

[33] M. Ferrero, C. Haetinger, Higher derivations of semiprime rings, *Comm. Algebra* **30** (5) (2002), 2321-2333.

[34] M. Gerstenhaber, S.D. Schack, Algebraic cohomology and deformation theory in deformation theory of algebras and structures and applications. *NATO ASI Ser. C* **247**, Kluver, Dordrecht , (1988), 11-264.

[35] J.S. Golan, Extension of derivations to modules of quotients, *Comm. Algenra* **9**(3) (1981), 275-281.

[36] J.S. Golan, Torsion Theory, *Pitman Monographs and Surveys in Pure and Applied Mathematics* Vol. 29, Longman Scientific & Technical, Harlow; John Wiley & Sons, New York, 1986.

[37] K.R. Goodearl, Global dimension of differential operator rings, *Proc. Amer. Math. Soc.* **45** (1974), 315-322.

[38] K.R. Goodearl, Global dimension of differential operator rings II, *Trans. Amer. Math. Soc.* **209** (1975), 65-85.

[39] K.R. Goodearl, Global dimension of differential operator rings III, *J. London Math. Soc.***(2)** 17 (1978), 397-409.

[40] K.R. Goodearl, T.H. Lenagan, Constructing bad noetherian local domains using derivations, *J. Algebra* **123** (1989), 478-495.

[41] C. Haetinger, *Derivações de ordem superior em anéis primos e semiprimos*, Porto Alegre, UFRGS (Brazil). PhD Thesis. Advisor: M. Ferrero, (2000), available at the IMUFRGS Digital Library, inscription number 000277135, http://www.biblioteca.ufrgs.br/bibliotecadigital.

[42] C. Haetinger, *Derivações em anéis primos e semiprimos, Porto Alegre, UFRGS (Brazil).* Master Dissertation. Advisor: M. Ferrero. (1993), available at the IMUFRGS Digital Library, http://www.biblioteca.ufrgs.br/bibliotecadigital, inscription number 000108957.

[43] C. Haetinger, Higher derivations on Lie ideals, *Série TEMAS: Tendências em Matemática Aplicada e Computacional*, São Paulo (Brazil), ISSN 8688-3026, volume 3, parte 1, número 1 (2002), 141-145.

[44] H. Hasse, Theorie der Differentiale in algebraischen Funktionenkörper mit vollkommenem Konstantenkörper, *Journal für Mathematik* **172** (1934), 55-64.

[45] H. Hasse, Theorie der höheren Differentiale in einem algebraischen Funktionenkörper mit vollkommenem Konstantenkörper bei beliebiger Charakteristik", *J. Reine Angew. Math. (Journal für Mathematik)* **175** (1936), 50-54.

[46] H. Hasse, Noch eine Begründung der Theorie der höheren Differentialquotienten in einem algebraischen Funktionenkörper einer Unbestimmten", *J. Reine Angew. Math. (Journal für Mathematik)* **177** (1936), 215-237.

[47] N. Heerema, Convergent higher derivations on local rings, *Trans. Amer. Math. Soc.* **132** (1968), 31-44.

[48] N. Heerema, Derivations and automorphisms of complete regular local rings, *Amer. J. Math.* **88** (1966), 33-42.

[49] N. Heerema, Derivations and embeddings of a field in its power series ring, *Proc. Amer. Math. Soc.* **11** (1960), 188-194.

[50] N. Heerema, Higher derivations and automorphisms of complete local rings, *Bull. Amer. Math. Soc.* **76** (1972), 1212-1225.

[51] N. Heerema, Inertial automorphisms of a class of wildly ramified v-rings, *Trans. Proc. Amer. Math. Soc.* **132** (1968), 45-54.

[52] I.N. Herstein, On the Lie structure of an associative ring, *J. Algebra* **14** (1970), 561-571.

[53] I.N. Herstein, *Topics in Ring Theory*, Univ. of Chicago Press, Chicago, 1969.

[54] I.N. Herstein, Jordan derivations of prime rings, *Proc. Amer. Math. Soc.* **8** (1957), 1104-1110.

[55] R.G. Heyneman, M.E. Sweedler, Affine Hopf Algebras I, *J. Algebra* **13** (1969), 192-241.

[56] N. Jacobson, Galois theory of purely inseparable fields of exponent one, *Amer. J. Math.* **66** (1944), 645-648.

[57] N. Jacobson, *Lectures in abstract algebra*, Vol. III, Van Nostrand, Princeton, New Jersey, 1964.

[58] N. Jacobson, Lie algebras, *Interscience Tracts in Pure and Appl. Math.*, vol. 10, Interscience, New York, 1962.

[59] A. Jaeger, Eine algebraische Theorie vertauschbarer Differentiationen für Körper beliebiger Charakteristik", *J. Reine Angew. Math. (Journal für Mathematik)* **190** (1952), 1-21.

[60] W. Jing, S. Lu, Generalized Jordan derivations on prime rings and standard operator algebras, *Taiwanese J. Math.* **7** (4) (2003), 605-613.

[61] S. Jøndrup, Automorphisms of upper triangular matrix rings, *Arch. Math.* **49** (1987), 497-502.

[62] Y.-S. Jung, Generalized Jordan triple higher derivations on prime rings, *Indian J. Pure & Appl. Math.* Vol. 36(9) (2005), 513-524.

[63] V.K. Kharchenko, Differential Identities of Prime Rings, *Algebra i Logika* **17**(2) (1978), 220–238; or *Algebra and Logic* **17** (1978), 155-168.

[64] M. Koppinen, Automorphisms and higher derivations of incidence algebras, *J. Algebra* **174** (1995), 698-723.

[65] M. Koppinen, Three automorphism theorems for triangular matrix algebras, *Linear Algebra Appl.* **245** (1996), 295-304.

[66] Y. Lequain, Differential simplicity and complete integral closure, *Pacific Journal of Mathematics* **36** (1971), 741-751.

[67] A. Leroy, J. Matczuk, Derivations et automorphismes algebriques d'anneux premiers, *Communications in Algebra* **13** (6) (1985), 1245-1266.

[68] D.R. Malm, Global and Krull dimension of Schmidt differential operator rings, *Comm. Algebra,* **17** (10) (1989), 2415-2459.

[69] D.R. Malm, Simplicity of partial and Schmidt differential operator rings, *Pacific Journal of Mathematics* **132** (1) (1988), 85-112.

[70] H. Matsumura, Integrable derivations, *Nagoya Math. J.* **87** (1982), 227-245.

[71] A. Nakajima, On categorical properties of generalized derivations, *Scientiae Mathematicae* **2** (3) (1999), 345-352.

[72] A. Nakajima, Generalized Jordan derivations, *Proceedings of the third Korea-China-Japan International Symposium on Ring Theory* (2000), 235-243.

[73] Y. Nakai, High order derivations I, *Osaka J. Math.* **7** (1) (1970), 1-27.

[74] Y. Nakai, On locally finite iterative higher derivations, *Osaka J. Math.* **15** (3) (1978), 655-662.

[75] A. Nowicki, Derivations of special subrings of matrix rings and regular graphs, *Tsukuba Journal of Mathematics* **7** (1983), 281-297.

[76] A. Nowicki, Higher *R*-derivations of special subrings of matrix rings, *Tsukuba Journal of Mathematics* **8** (1984), 227-253.

[77] A. Nowicki, Inner derivations of higher orders, *Tsukuba Journal of Mathematics* **8** (2) (1984), 219-225.

[78] H. Osborn, Modules of differentials I, *Math. Ann.* **170** (1967), 221-244.

[79] A. Ouarit, Dérivations et automorphismes algébriques d'anneaux semi-premiers, *C. R. Acad. Sci. Paris* **314** (1) (1992), 241-244.

[80] G. Paulus, *Inseparabilität und Differentiation in Körpern der Charakteristik p*, Zulassungsarbeit, Heidelberg, 1954.

[81] E.C. Posner, Derivations in prime rings, *Proc. Amer. Math. Soc.* **8** (1957), 1093-1100.

[82] P. Ribenboim, Higher derivations of rings I, *Revue Roumaine de Math. Pures et Appl.* **16** (1971), 77-110.

[83] P. Ribenboim, Higher derivations of rings II, *Revue Roumaine de Math. Pures et Appl.* **16** (1971), 245-272.

[84] P. Ribenboim, Higher order derivations of modules, *Portugaliae Mathematica* **39** (1-4) (1980), 381-397.

[85] S.H. Rim, Extension of higher anti-derivations to modules of quotients, *J. Korean Math. Soc.* **24**(1) (1987), 25-31.

[86] A. Seidenberg, Derivations and integral closure, *Pacific Journal of Mathematics* **16** (1966), 167-173.

[87] A. Seidenberg, Differential ideals in complete local rings, *Amer. J. Math.* **95** (1973), 52-58.

[88] A. Seidenberg, Differential ideals in rings of finitely type, *Amer. J. Math.* **89** (1967), 22-42.

[89] B. Singh, Higher derivations, normal flatness and analytic products, *J. Algebra* **95** (1985), 236-244.

[90] P. Šemri, Additive derivations of some operator algebras, *Illinois Journal of Mathematics,* **35** (1991), 234-240.

[91] M.E. Sweedler, Hopf algebras with one group-like element, *Trans. Amer. Math. Soc.* **127** (1967), 515-526.

[92] M.E. Sweedler, Structure of inseparable extensions, *Ann. of Math.* **(2)** 87 (1968), 401-410.

[93] O. Teichmüller, Differentialrechnung bei Charakteristik *p*, *J. Reine Angew. Math. (Journal für Mathematik)* **175** (1936), 89-99.

[94] O. Teichmüller, Über die Struktur diskret bewerterer perfekter Körper, *Gott. Nach.* vol. 107 (1933), 587-623.

[95] R. Vidal, Anneaux de valuation discrète complets non commutatifs, *Trans. Amer. Math. Soc.* **267** (1) (1981), 65-81.

[96] A.R. Villena, Lie derivations on Banach algebras, *J. Algebra* **226** (2000), 390-409.

[97] E.R. Voss, On the isomorphism problem for incidence rings, *Illinois Journal of Mathematics* **24** (1980), 624-628.

[98] M. Weisfeld, Purely inseparable extensions and higher derivations, *Trans. Amer. Math. Soc.* **116** (1965), 435-449.

[99] M. Zariski, Studies in equisingularity I, *Amer. J. Math.* **87** (1965), 507-536.

[100] F. Zerla, Iterated higher derivation in fields of prime characteristic, *Michigan Math. J.* **15** (1968), 407-415.

In: Mathematical Theories, Models and Groups ISBN 978-1-61324-250-6
Editor: Marcus J. Grecho

Chapter 4

PEER GROUP SITUATIONS AND GAMES WITH INTERVAL UNCERTAINTY

***Rodica M. Branzei*[1], *Lina Mallozzi*[2] *and Stef H. Tijs*[3]**
[1] Faculty of Computer Science
"Alexandru Ioan Cuza" University, Iaşi, Romania
[2]Dipartimento di Matematica e Applicazioni R.Caccioppoli
Napoli, Italy
[3] CentER and Department of Econometrics and Operations Research
Tilburg University, Tilburg, Netherlands

1. Introduction

Several papers about cooperative interval-valued games have appeared recently ([1–5]). In this case a cooperative game is considered with an interval-valued characteristic function, i.e. the worth of a coalition is not a real number, but a compact interval of real numbers. This means that one observes a lower bound and an upper bound of the worth of the considered coalitions. This is very important, for example from a computational and algorithmic point of view.

Let us now recall basic definitions of interval analysis ([9]). Let IR be the set of real intervals $\mathrm{IR} = \{[\underline{I}, \overline{I}] \subset \mathcal{R}, \underline{I}, \overline{I} \in \mathcal{R}, \underline{I} \leq \overline{I}\}$. For any $I \in \mathrm{IR}$, the values $\underline{I}, \overline{I}$ are called lower and upper bound respectively of the interval I. Two intervals are equal if their respective bounds coincide. For any real number $k \in \mathcal{R}$, we also denote $k = [k,k]$.

Let $I, J \in \mathrm{IR}$; we consider the following partial order $I \succcurlyeq J$ iff $\underline{I} \geq \underline{J}$ and $\overline{I} \geq \overline{J}$. If $I = [a,a]$, $a \in \mathcal{R}$ and $J = [b,b]$, $b \in \mathcal{R}$, we have that $I \succcurlyeq J \Longleftrightarrow a \geq b$.

Remark that $\succcurlyeq$ is a partial order relation in IR: for example $[0,2] \succcurlyeq [-1,2]$, but $[-2,3]$ and $[-1,2]$ are not comparable with respect to $\succcurlyeq$.

The sum operation is defined in IR as follows:

$$I+J=[\underline{I}+\underline{J},\overline{I}+\overline{J}].$$

The product of an interval $I \in$ IR times a non-negative real number $\alpha \geq 0$ is defined as follows:

$$\alpha I=[\alpha\underline{I},\alpha\overline{I}].$$

We denote by IR_+ the set of intervals $I \succcurlyeq [0,0]$ and by IR^n_+ the cartesian product $\mathrm{IR}^n_+ = \{(A_1,...,A_n), A_i \in \mathrm{IR}_+, i=1,...,n\}$.

A cooperative interval game is an ordered pair $\langle N,w\rangle$, where $N=\{1,...,n\}$ is the set of the players and $w: 2^N \to$ IR is the characteristic function such that $w(\emptyset)=[0,0]$. We denote by $w(S)=[\underline{w}(S),\overline{w}(S)]$ the worth of the coalition S. The border games $\langle N,\underline{w}\rangle$, $\langle N,\overline{w}\rangle$ and the length game $\langle N,|w|\rangle$ are classical TU-games with characteristic functions $\underline{w}$, $\overline{w}$ and $|w|=\overline{w}-\underline{w}$ respectively.

A cooperative interval game $\langle N,w\rangle$ is convex if it is supermodular, i.e.

$$w(S\cup T)+w(S\cap T)\succcurlyeq w(S)+w(T),\ \forall S,T\in 2^N,$$

and the length game $\langle N,|w|\rangle$ is also supermodular (convex) in the classical definition, i. e.

$$|w|(S\cup T)+|w|(S\cap T)\geq |w|(S)+|w|(T),\ \forall S,T\in 2^N.$$

A cooperative interval game $\langle N,w\rangle$ is size monotonic if $\langle N,|w|\rangle$ is monotonic, i.e. $|w|(S)\leq |w|(T)$ for any $S,T\in 2^N$ with $S\subset T$.

If $|I|\geq |J|$ we define the difference interval $I-J=[\underline{I}-\underline{J},\overline{I}-\overline{J}]$.

The definition of interval-valued cooperative games has been given in [5], together with several notions of balancedness and cores by using selections of such games. In [1–4] the authors investigate several interval solutions on the class of interval-valued games and convexity properties of such games. They also give the definitions of Shapley value and Weber set on the class of size monotonic interval games which includes the class of convex interval games.

In this paper, by using the introduced notions, we approach the study of peer group games in an interval-valued setting. Peer group games have been introduced in [6] and correspond to situations where the players have a hierarchical relationship and a positive real value is given to each player only if all his superiors cooperate with him. Peer group games are a subclass of the class of trading games ([7]) and also of the class of games with permission structure ([8]). We introduce interval peer group situations and for the related interval peer group games we give properties of monotonicity, convexity and specify the Shapley value and the Weber set. In the last section all the results are applied to sealed bid second price auctions with uncertainty.

2. Interval Peer Group Situations and Games

We consider a finite set of agents $N = \{1,2,...,n\}$ having a hierarchical relationship given by a directed graph $\mathcal{T}$ where each agent is located in a node in such a way that there is one node as root corresponding to agent 1 and there exists a unique directed path from each node to the root (Fig. 1). If all the nodes lie on a single directed path we have a chain.

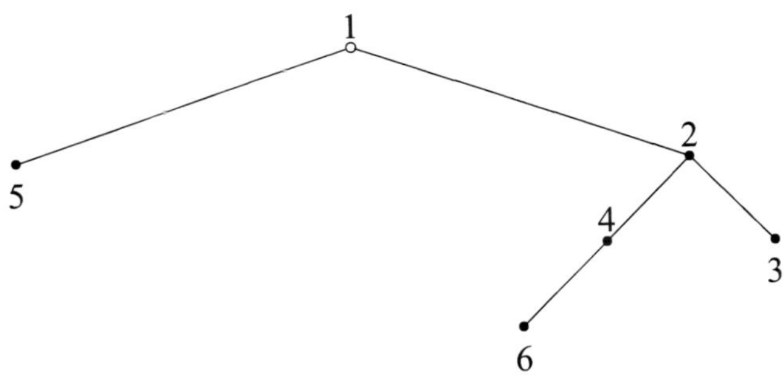

Figure 1.

In [6] peer group situations and then peer group games have been defined: a positive gain is associated to each agent only in the case where all his superiors cooperate with him. Recall that a triplet $\langle N,P,a\rangle$ is a $\mathcal{T}$**-connected peer group situation** where:

i) N is the set of involved agents;

ii) $P : N \to 2^N$ with $P(i) = [1,i]$ mapping to any agent i the set of all agents in the path of $\mathcal{T}$ connecting 1 to i denoted by $[1,i]$;

iii) $a \in \mathbb{R}^n_+$ is the vector of agents' potential economic possibilities.

In the directed graph in Fig. 1 we have $N = \{1,2,3,4,5,6\}$ and $[1,1] = \{1\}, [1,2] = \{1,2\}, [1,3] = \{1,2,3\}, [1,4] = \{1,2,4\}, [1,5] = \{1,5\}, [1,6] = \{1,2,4,6\}$.

A **peer group game** $\langle N, v_{P,a}\rangle$ or shortly $\langle N, v\rangle$ is defined with $N = \{1,2,...,n\}$ and v given by

$$v(\emptyset) = 0;$$

$$v(S) = \sum_{i:P(i)\subseteq S} a_i, \qquad \text{for each } S \subseteq N.$$

Whenever $1 \notin S$, then $v(S) = 0$. Now we approach this situations in an interval-valued games setting. Namely we suppose that each agent i does not know exactly how much he gains from the cooperation with his superiors $[1,i]$; he knows only a lower and an upper bound of his gain given by a positive real interval $A_i = [\underline{A}_i, \overline{A}_i]$.

In this case we call an **interval $\mathcal{T}$-connected peer group situation** or shortly **ipg-situation** any triplet $\langle N, P, A\rangle$, where N, P are as in i), ii) respectively, and $A \in \text{IR}^n_+$ is the vector of agents' gain intervals.

Definition 2.1. *An* **interval-valued peer group game (ipg-game)** *corresponding to a ipg-situation $\langle N, P, A\rangle$ is an interval-valued cooperative game $\langle N, w_{P,A}\rangle$ or shortly $\langle N, w\rangle$ defined by $N = \{1,2,...,n\}$ and w given by*

$$w(\emptyset) = [0,0];$$

$$w(S) = \sum_{i:P(i)\subseteq S} A_i, \qquad \text{for each } S \subseteq N.$$

Whenever $1 \notin S$, then $w(S) = [0,0]$.

Example 2.1. [Sequential Production Situations] *We consider a production situation with n departments involved in the working process of a raw material. There is a hierarchy between them: the material is processed at stage k only after the processes in stages* 1,..., $k-1$. *At any stage there is a fixed cost necessary to process the material. Sometimes the cost may have an additional amount, for example due to a machinery accident. Let us suppose that the cost at stage k is in between $\underline{A}_k$ and $\overline{A}_k$. If we consider $N = \{1,...,n\}$, $P : N \to 2^N$ s.t. $P(k) = [1,k]$ and $w(S) = \sum_{i:P(i)\subseteq S} A_i$,* for each $S \subseteq N$, *the game $\langle N, w_{P,A}\rangle$ is an interval peer group game. Note that $w(S) = 0$ for any coalition S not containing the root* 1.

For example, let us consider $N = \{1,2,3\}$ and $A_i = [a_i, a_i]$, $a_i \in \mathbb{R}$, $i = 1,3$, $A_2 = [\underline{A}_2, \overline{A}_2]$. We have

$$w(\{1\}) = w(\{1,3\}) = [a_1, a_1],$$

$$w(\{1,2\}) = [a_1 + \underline{A}_2, a_1 + \overline{A}_2],$$
$$w(\{1,2,3\}) = [a_1 + \underline{A}_2 + a_3, a_1 + \overline{A}_2 + a_3],$$

and $w(S) = [0,0]$ *in any other case. The uncertainty due to department* 2 *affects the departments that are not its superiors. In a non-cooperative setting, the production situations have been studied in [11].*

We recall the definition of unanimity interval-valued games ([6]). Given $J \in IR^n_+$ and $T \in 2^N \setminus \{0\}$, the unanimity game corresponding to T and J is defined by

$$u_{T,J}(S) = \begin{cases} \text{J}, & T \subseteq S, \\ [0,0], & \text{otherwise}, \end{cases}$$

for each $S \in 2^N$. We notice that $v_{T,J} = Ju_T$.

Any interval-valued peer group game can be expressed in terms of unanimity interval-valued games as specified in the following.

Proposition 2.1. *Interval-valued peer group games with fixed N, P form a cone*

$$C^{ipg} = \{\langle N, w_{P,A}\rangle,\ A \in IR^n_+\}.$$

Moreover for any $\langle N, P, w_{P,A}\rangle \in C^{ipg}$ we have

$$w_{P,A} = \sum_{i=1}^{n} u_{[1,i],A_i}.$$

Proof. For any $\alpha, \beta \in \mathbb{R}_+$ and $A, B \in IR^n_+$, given the hierarchy $\mathcal{T}$ described by P, we have

$$\alpha w_{P,A} + \beta w_{P,B} = w_{P,\alpha A + \beta B}.$$

In fact for any $S \in 2^N$ we have

$$\alpha \sum_{i:P(i)\subseteq S} A_i + \beta \sum_{i:P(i)\subseteq S} B_i = \sum_{i:P(i)\subseteq S} (\alpha A_i + \beta B_i).$$

Moreover for each $S \in 2^N$ by using the definition of unanimity interval-valued games we have

$$w_{P,A}(S) = \sum_{i=1}^{n} u_{[1,i],A_i}(S).$$

■

Some solution concepts, well-known in the theory of cooperative interval games, have special properties for interval peer group games as specified below.

Proposition 2.2. *Let* $\langle N, P, w_{P,A}\rangle$, $A \in IR^n_+$ *be an interval-valued peer group game. Then:*
1. $\langle N, P, w_{P,A}\rangle$ *is size monotonic;*
2. $\langle N, P, w_{P,A}\rangle$ *is convex;*
3. the interval Shapley value $\Phi_i(w) = \sum_{j:i\in P(j)} \frac{A_j}{|P(j)|}$, $i \in N$, *where* $|P(j)|$ *is the number of elements in* $P(j)$;
4. the interval Weber set is contained in the interval core: $\emptyset \neq \mathcal{W}(w) \subset \mathcal{C}(w)$, *where* $\mathcal{W}(w) = conv\{m^\sigma(w) \, : \, \sigma \in \Pi(N)\}$, $m^\sigma(w)$ *being the interval marginal contribution vector of* w *with respect to the permutation* σ.

Proof. 1. Let us consider $S, T \in 2^N$ with $S \subset T$. For any $i \in N$ such that $[1,i] \subset S \subset T$ we have $u_{[1,i],A_i}(S) = A_i = u_{[1,i],A_i}(T)$, and for any $i \in N$ such that $[1,i] \subset T$ and $[1,i]$ not subset of S we have $u_{[1,i],A_i}(T) = A_i \succcurlyeq u_{[1,i],A_i}(S) = [0,0]$. By considering the sum for $i = 1,...,n$ we have $w_{P,A}(T) \succcurlyeq w_{P,A}(S)$. Then $|w_{P,A}|(T) = \sum_{i=1}^{n} (\overline{A}_i - \underline{A}_i) \geq 0 = |w_{P,A}|(S)$.

2. As remarked in Section 3 of [2], unanimity interval games are convex interval games. By using Proposition 2.1 we have the result.

3. By using Proposition 4.3 in [2] the interval Shapley value is the interval $\Phi_i(w) = [\Phi_i(\underline{w}), \Phi_i(\overline{w})]$; from Proposition 1 in [6] the Shapley value of a peer group game $\langle N, P, v_{P,a}\rangle$, $a \in \mathbb{R}^n_+$ is the vector $\Phi_i(v) = \sum_{j:i\in P(j)} \frac{a_j}{|P(j)|}$, $i \in N$.

4. By using Theorem 4.1 and Proposition 4.1 in [2]. ■

3. Sealed Bid Second Price Auctions with Uncertainty

A seller of an object decides a reservation price r as the lowest price for which he wants to sell the object and it is known to the bidders ([10]). There are n bidders $\{1, 2, ..., n\}$, the players, that submit a secret bid: the bidder with the highest bid obtains the object at the price of the second highest bid. We suppose that $w_i \geq r > 0$ for $i = 1, ..., n$ is the value for player i of the object, and that

$$w_1 > w_2 > w_3 > .....w_n \geq r.$$

The value w_i is known only by player i.

Now, we suppose that the bid w_i of player i is not exactly determined: the bidder submits an interval of values $W_i = [w_i - \delta_i, w_i + \delta_i]$ where $\delta_i, i = 1, ..., n$ represents bidder ith's uncertainty. Here the bidder ith's value can be any element in $W_i = [w_i - \delta_i, w_i + \delta_i]$. These value bounds are not known to the other players. In our model we suppose that the bidder with the highest bid obtains the object at the highest interval price of the second highest bid with respect to the operator $\succcurlyeq$ in IR.

Suppose that for the given $\delta_1, ..., \delta_n$, the following assumptions hold:

$$W_1 \succcurlyeq W_2 \succcurlyeq W_3 \succcurlyeq .....W_n \succcurlyeq [r, r], \tag{3.1}$$

$$|W_1| > |W_2| > |W_3| > .....|W_n| > 0. \tag{3.2}$$

Note that $W_n = [\underline{w}_n, \overline{w}_n] \succcurlyeq [r, r]$ means $\underline{w}_n \geq r$. Such $\delta_i, i = 1, ..., n$ exist. For example, if we consider

$$\delta_i = \frac{w_i - r}{n}, i = 1, ..., n$$

we have

$$w_{i+1} - \delta_{i+1} \leq w_i - \delta_i \Leftrightarrow (w_i - w_{i+1}) \left(1 - \frac{1}{n}\right) \geq 0$$

and also

$$w_{i+1} + \delta_{i+1} \leq w_i + \delta_i \Leftrightarrow (w_{i+1} - w_i) \left(1 + \frac{1}{n}\right) \leq 0$$

so that $W_i \succcurlyeq W_{i+1}$ for all $i = 1, ..., n-1$, implying that assumption (3.1) holds. Moreover,

$$2\frac{w_{i+1} - r}{n} \leq 2\frac{w_i - r}{n} \Leftrightarrow w_{i+1} \leq w_i$$

so that $|W_i| > |W_{i+1}|$ for all $i = 1, ..., n-1$, implying that assumption (3.2) holds.

For any possible coalition $S \subseteq \{1, 2, ..., n\}$, let us define the payoff as follows.

i) $S = \{i\}$: player i bids W_i that gives to him a payoff $w(\{i\}) = [0, 0]$ if $i \neq 1$, $w(\{1\}) = W_1 - W_2$ because player 1 obtains the object at price in W_2;

ii) $S = N$, all the players cooperate: optimal bid for player 1 is to choose in W_1, for the others is to choose $[r, r]$; player 1 obtains the object at price r and the payoff to N is $w(N) = W_1 - [r, r]$;
iii) $S \subset N$, let $i(S)$ be the player with the highest (interval) value with respect to the operator $\succcurlyeq$ in IR:

- if $1 \notin S$ then $i(S)$ bids $W_{i(S)}$, the others bid r. The object goes to player 1 and the value of coalition S is $w(S) = [0, 0]$;

- if $1 \in S$ then the highest bid is in W_1 and the second highest is in W_{k+1} if $[1, k] \subset S$ and $k+1 \notin S$, since $W_i = [r, r]$ for $i = 2, ..., k$. In this case the value of the coalition is $w(S) = W_1 - W_{k+1}$.

Let us define $\mathcal{T}$ the line graph with root 1 and arcs $(i, i+1)$, $i = 1, ..., n-1$, and the hierarchy function $P : N \to 2^N$, $P(i) = \{1, ..., i\}$. Let us define $A_i = W_i - W_{i+1}$ for any $i \in N$: the triplet $\langle N, P, A\rangle$ with $A \in \mathrm{IR}^n_+$ is an interval peer group situation.

The **auction peer group interval game** is the cooperative interval game $\langle N, w_{P,A}\rangle$ or $\langle N, w\rangle$ where

$$w_{P,A}(S) = \sum_{i=1}^{n} u_{[1,i],W_i - W_{i+1}}(S) \quad \text{for each } S \in 2^N,$$

where $W_{n+1} = [r, r]$. In fact for any $S \subseteq N$ we have: $w(S) = [0, 0]$ if $1 \notin S$, $w(S) = \sum_{i=1}^{k}(W_i - W_{i+1}) = W_1 - W_{k+1}$ if $[1, k] \subset S$ and $k+1 \notin S$.

Example 3.1. *As in Example 4 in [6], we consider three bidders in an auction have values for the object of 100, 80, 50, respectively, and the reservation price is $r = 25$. Let us suppose that the three bidders are facing interval uncertainty with $\delta_i = \frac{w_i - r}{3}$, $i = 1, 2, 3$. We have*

$$W_1 = [100 - 25, 100 + 25],\ W_2 = \left[80 - \frac{55}{3}, 80 + \frac{55}{3}\right],\ W_3 = \left[50 - \frac{25}{3}, 50 + \frac{25}{3}\right].$$

Let us define the auction peer group interval game $\langle N, w\rangle$ where $N = \{1, 2, 3\}$ and the characteristic function is

$$w(\{1\}) = W_1 - W_2 = \left[\frac{115}{3} - 25, \frac{115}{3} + 25\right],$$

$$w(\{1,2\}) = W_1 - W_3 = \left[\frac{175}{3} - 25, \frac{175}{3} + 25\right],$$

$$w(\{1,2,3\}) = W_1 - r = [75 - 25, 75 + 25],$$

and for any other coalition $S \subseteq N$, we have $w(S) = [0,0]$.

References

[1] S.Z. Alparslan Gök, R. Branzei, S. Tijs, *Cores and stable sets for interval-valued games*, DP No. 2008-17, CentER Tilburg University, 2008.

[2] S.Z. Alparslan Gök, R. Branzei, S. Tijs, Convex interval games, *J. Appl. Math. Dec. Sc.* (2009), Article ID 342089, 14 pages, DOI: 10.1115/2009/342089.

[3] S.Z. Alparslan Gök, R. Branzei, S. Tijs, Airport interval games and their Shapley value, *Oper. Res. Dec.* **2** (2009), 571–575.

[4] S.Z. Alparslan Gök, R. Branzei, S. Tijs, *The interval Shapley value: an axiomatization*, Preprint no. 130, Institute of Applied Mathematics METU, to appear in *Cent. Eur. J. Oper. Res.* (2009), DOI: 10.1007/s10100-009-0096-0.

[5] S.Z. Alparslan Gök, S. Miquel, S. Tijs, Cooperation under interval uncertainty, *Math. Meth. Oper. Res.* **69** (2009), 99–109.

[6] R. Branzei, V. Fragnelli, S. Tijs, Tree-connected peer group situations and peer group games, *Math. Meth. Oper. Res.* **25** (2002), 93–106.

[7] X.T. Deng, C. Papadimitriou, On the complexity of cooperative solution concepts, *Math. Oper. Res.* **19** (1994), 257–266.

[8] R.P. Gilles, G. Owen, R. van den Brink, Games with permission structures: the conjunctive approach, *Inter. J. Game Th.* **20** (1992), 277–293.

[9] R. Moore, Methods and applications of interval analysis, *SIAM St. Appl. Math.*, 1995.

[10] S. Tijs, *Introduction to game theory*, Hindustan Book Agency, India, 2003.

[11] M. Voorneveld, S. Tijs, L. Mallozzi, Sequential production situations and potentials, In: Patrone F, Garcia-Jurado I, Tijs S (Eds), *Game Practice: Contributions from Applied Game Theory, Theory and Decision Library C* **23** (1999), Kluwer Academic Publishers: Boston, 241–258.

In: Mathematical Theories, Models and Groups
Editor: Marcus J. Grecho
ISBN 978-1-61324-250-6

Chapter 5

ON SHIFT-INVARIANT SOLUTIONS FOR TU-GAMES

Kensaku Kikuta
School of Business Administration
University of Hyogo
Kobe, Japan

1. Introduction

In a TU-game any player should not take more from the game than his maximal incremental contribution to any coalition. A payoff that exceeds this contribution may be regarded unreasonable for the player. The reasonable set of a cooperative game is the set of all payoff vectors that are reasonable in this sense, and it was first introduced by Milnor [5]. Adding a concept of minimal incremental contribution, Gerard-Varet and Zamir [1] defined a set of reasonable outcomes and justified it by axiomatization. These sets as well as the imputation set are considered as pre-solution-concepts. It is asserted not that outcomes within the sets are necessarily plausible, but only that those outside the sets are implausible.

Since the reasonable set and the set of reasonable outcomes are large, it would be preferable if we could define a subset of them and if it preserves any desirable properties as many as possible. A decision-maker would get a more precise guideline if he/she has a smaller set as a pre-solution. Milnor [5] defined two other pre-solutions, known as "L" and "D," which put lower and upper bounds on the payoff to any coalition. Tijs [9] studied a hypercube, which is a core cover when the core is not empty, and it is defined by using the marginal contribution of any player for the grand coalition. Llerena and Rafels [2] axiomatizes the core, where one of the axioms is expressed by minimal and maximal marginal contributions.

Shapley and Shubik [8] introduced the shifted game to define quasi-cores. The term "shifted" appeared in Maschler, Peleg and Shapley [4] when it studies on a geometric property of the kernel and other solutions for TU games. Maschler, Peleg and Shapley [4] showed that a combination of making the shifted game and another operation for a game tends to shrink the reasonable set.

In this paper by an invariance-property of solutions for shifted games, we define a subset of the reasonable set of outcomes and characterize it. This set includes the prekernel and the least core, so it has a non-empty intersection with the core when the core is non-empty. Furthermore we characterize the set of pairwise reasonable outcomes.

An n-person cooperative game with side payments (abbreviated as a *game*) is an ordered pair (N,v), where $N=\{1,2,\ldots,n\}$ is the set of *players* and v, called the *characteristic function*, is a real-valued function on the power set of N, satisfying $v(\emptyset)=0$. We denote by G_N the set of all games in which N is the set of players. For simplicity we express a game $(N,v)\in G_N$ as v. A subset of N is called a *coalition*. For any set Z, $|Z|$ denotes the cardinality of Z. For a coalition S, R^S is the $|S|$-dimensional product space $R^{|S|}$ with coordinates indexed by players in S. For $x\in R^S$, the component for $i\in S$ is denoted by x_i. For $S\subseteq N$ and $x\in R^N$, we define $x(S)=\sum_{i\in S} x_i$ (if $S\neq\emptyset$) and $=0$ (if $S=\emptyset$). For $a,b\in R$, $[a,b]$ is an interval in R. For $a,b\in R^N$, $[a,b]\equiv\prod_{i\in N}[a_i,b_i]$ is an interval in R^N. For a set $Y\subseteq R^N$, we let $\ell_i(Y)\equiv\inf_{y\in Y}\{y_i\}$ and $u_i(Y)\equiv\sup_{y\in Y}\{y_i\}$ for every $i\in N$. For a set $Y\subseteq R^N$ and a positive number $k>0$, we let $kY\equiv\{ky:y\in Y\}$. For a set $Y\subseteq R^N$ and $x\in R^N$, we let $x+Y\equiv\{x+y:y\in Y\}$. For a real number $\varepsilon\in R$, we define the (ε-)*shifted* game of a game $v\in G_N$, denoted by v_ε, by

$$v_\varepsilon(S)=\begin{cases} v(S)-\varepsilon, & \text{if } S\neq N,\emptyset;\\ v(S), & \text{if } S=N,\emptyset.\end{cases}$$

A *solution* is a mapping ϕ which associates with any game $v\in G_N$ a subset $\phi(v)$ of R^N. Here we define four solutions,[1] three of which are called

[1] In this paper each solution is expressed by uppercase letters in a calligraphic style, as well

pre-solutions and they appear in later sections[2]. Let $M_i(v) \equiv \max_{i \in S}\{v(S) - v(S\backslash\{i\})\}$ and $m_i(v) \equiv \min_{i \in S}\{v(S) - v(S\backslash\{i\})\}$ for all $i \in N$.

$$\mathcal{PI}(v) = \{x \in R^N : x(N) = v(N)\},$$
$$\mathcal{I}(v) = \{x \in R^N : x(N) = v(N) \text{ and } x_i \geq v(\{i\}), \forall i \in N\},$$
$$\mathcal{R}(v) = [m(v), M(v)],$$
$$\mathcal{E}(v) = \{(v(\{1\}) + \frac{v(N) - \sum_{i \in N} v(\{i\})}{n}, \ldots, v(\{n\}) + \frac{v(N) - \sum_{i \in N} v(\{i\})}{n})\}.$$

Some basic and well-known properties for solutions for TU-games are the following. These properties and their variations were used to characterize solutions for TU-games, for example, the nucleolus and the kernel.

Boundedness (**BDD**): For all $v \in G_N$, $\phi(v)$ is a bounded set in R^N.

Covariance (**COV**): For all $v \in G_N$, all $k > 0$, all $y \in R^N$, if w is such that $w(S) = kv(S) + y(S)$ for every $S \subseteq N$, then $\phi(w) = k\phi(v) + y$.

Monotonicity (**MON**)[3] : For all $v, w \in G_N$, all $i \in N$, if $v(S) \geq w(S)$ for every $S \subseteq N$ then $\ell_i(\phi(v)) \geq \ell_i(\phi(w))$ and $u_i(\phi(v)) \geq u_i(\phi(w))$.

The next two properties are used with respect to our main result. The first property is well-known[4] .

Reasonableness (**RE**): For all $v \in G_N, \phi(v) \subseteq \mathcal{R}(v)$.

The second property says that a solution associates essentially the same set with a game and its shifted games. In Section 3 we will see examples of shift-invariant solutions and non-shift-invariant solutions.

Definition 1.1. *A solution ϕ is said to satisfy shift-invariance* (**SI**) *if : For all $v \in G_N$, all $\varepsilon \in R$, all $i \in N$, $\ell_i(\phi(v)) = \ell_i(\phi(v_\varepsilon))$ and $u_i(\phi(v)) = u_i(\phi(v_\varepsilon))$.*

---

as ϕ.

[2] Vectors in $\mathcal{PI}(v), \mathcal{I}(v), \mathcal{R}(v)$ are respectively called pre-imputations, imputations and reasonable outcomes.

[3] About discussions on monotonicity properties, see Sections 2.3 and 8.2 of Peleg and Sudhölter [7].

[4] For example, see Section 2.3 of Peleg and Sudhölter [7] and Llerena and Rafels [2].

Remark 1.1. *It is not difficult to show that the properties* **BDD**, **COV**, **MON** *and* **SI** *are independent. For example, $\mathcal{PI}$ satisfies* **COV**, **MON** *and* **SI**, *but does not satisfy* **BDD**. *The solution ϕ^1 defined by $\phi^1(v) = \{(\frac{v(N)}{n}, \ldots, \frac{v(N)}{n})\}$ satisfies* **BDD**, **MON** *and* **SI**, *but it does not satisfy* **COV**. *The solution ϕ^2 defined by $\phi^2(v) = \{(v(\{1\}), \ldots, v(\{n\}))\}$ satisfies* **BDD**, **MON** *and* **COV**, *but it does not satisfy* **SI**. *The solution ϕ^3 defined by $\phi^3(v) = \{(\frac{v(\{1\})+v(N)-v(N\setminus\{1\})}{2}, \ldots, \frac{v(\{n\})+v(N)-v(N\setminus\{n\})}{2})\}$ satisfies* **BDD**, **SI** *and* **COV**, *but it does not satisfy* **MON**. *Therefore these four properties are independent.*

Furthermore the next proposition relates four properties above.

Proposition 1.1. *Assume a solution ϕ satisfies* **BDD**, **COV** *and* **MON**. *Furthermore, assume that $\phi(v) \neq \emptyset$ for all $v \in G_N$. Then it satisfies* **RE**.

Proof. Assume a solution ϕ satisfies **BDD**, **COV** and **MON**. Let o be a game such that $o(S) = 0$ for every $S \subseteq N$. Since ϕ satisfies **BDD**, $\phi(o)$ is bounded, and so there exist $a, b \in R^N$ such that $\phi(o) \subset [a, b]$. By the assumption, $\phi(o) \neq \emptyset$. Assume $x \in \phi(o)$ and $x \neq 0$. Take a positive number $k > 0$ such that $kx \notin [a, b]$. For this k, define a game o' by $o'(S) = ko(S)$ for every $S \subseteq N$. Since $o'(S) = o(S)$ for every $S \subseteq N$ we have $\phi(o) = \phi(o') = k\phi(o)$ by **COV**. Since $x \in \phi(o)$ we must have $kx \in k\phi(o) = \phi(o)$. This contradicts $\phi(o) \subset [a, b]$. So $x \in \phi(o)$ implies $x = 0$. Hence we have $\phi(o) \subseteq \{0\}$. By the assumption of the proposition, $\phi(o) \neq \emptyset$. Hence we must have $\phi(o) = \{0\}$. For each $x \in R^N$, let w_x be a game such that $w_x(S) = o(S) + x(S)$ for every $S \subseteq N$. Then by **COV**, $\phi(w_x) = \phi(o) + x = \{0\} + x = \{x\}$. That is,

$$\phi(w_x) = \{x\}, \text{ for all } x \in R^N. \tag{1.1}$$

Let $v \in G_N$. In (1.1), letting $x = m(v)$ and $x = M(v)$, we have respectively $\phi(w_{m(v)}) = \{m(v)\}$ and $\phi(w_{M(v)}) = \{M(v)\}$. For every $S \subseteq N$, let $S = \{i_1, \ldots, i_s\}$, and $s = |S|$. Then

$$v(S) = \sum_{\alpha=2}^{s} [v(\{i_1, \ldots, i_\alpha\}) - v(\{i_1, \ldots, i_{\alpha-1}\})] + v(\{i_1\})$$
$$\begin{cases} \geq \sum_{\alpha=2}^{s} m_{i_\alpha}(v) + m_{i_1}(v) = \sum_{i \in S} m_i(v) = w_{m(v)}(S), \\ \\ \leq \sum_{\alpha=2}^{s} M_{i_\alpha}(v) + M_{i_1}(v) = \sum_{i \in S} M_i(v) = w_{M(v)}(S). \end{cases}$$

Hence by **MON**, for every $i \in N$, $\ell_i(\phi(v)) \geq \ell_i(\phi(w_{m(v)})) = m_i(v)$ and $u_i(\phi(v)) \leq u_i(\phi(w_{M(v)})) = M_i(v)$. So $\phi(v) \subseteq \mathcal{R}(v)$. ■

2. A Refinement of the Reasonable Set

In this section we characterize a largest-under-inclusion solution with the property of shift-invariance. Define a solution $\mathcal{S}$ by: For $v \in G_N$,

$$\mathcal{S}(v) \equiv \bigcap_{\varepsilon \in R} \mathcal{R}(v_\varepsilon).$$

If a solution ϕ satisfies **RE** and $\phi(v) = \phi(v_\varepsilon)$ for all $v \in G_N$ and all $\varepsilon \in R$ then it satisfies $\phi(v) \subseteq \mathcal{S}(v)$ for all $v \in G_N$. By the definition, $\mathcal{S}(v) \subseteq \mathcal{R}(v_0) = \mathcal{R}(v)$ for all $v \in G_N$, that is, $\mathcal{S}$ satisfies **RE**. The next example is to show that, in general, $\mathcal{S}(v) \neq \mathcal{R}(v)$.

Example 2.1. *Let* $N = 1,2,3$ $v(\{i\}) = 0$ *for all* $i \in N$ *and* $v(\{1,2\}) = 2, v(\{1,3\}) = 1, v(\{2,3\}) = 5, v(\{1,2,3\}) = 6$. *By definition*

$$M_1(v_\varepsilon) = \max\{1+\varepsilon, 2, -\varepsilon\}, M_2(v_\varepsilon) = \max\{5+\varepsilon, 5, -\varepsilon\}, M_3(v_\varepsilon) = \max\{4+\varepsilon, 5, -\varepsilon\},$$
$$m_1(v_\varepsilon) = \min\{1+\varepsilon, 1, -\varepsilon\}, m_2(v_\varepsilon) = \min\{5+\varepsilon, 2, -\varepsilon\}, m_3(v_\varepsilon) = \min\{4+\varepsilon, 1, -\varepsilon\},$$

and so

$$\mathcal{R}(v) = \mathcal{R}(v_0) = \{x \in R^3 : 0 \leq x_1 \leq 2, 0 \leq x_2 \leq 5, 0 \leq x_3 \leq 5\}.$$

It is easy to see that $(0,5,1) \in \mathcal{R}(v)$ *and* $(0,5,1) \notin \mathcal{R}(v_{-\frac{1}{2}})$ *since* $0 < m_1(v_{-\frac{1}{2}})$. *So* $\mathcal{S}$ *is a refinement of* $\mathcal{R}$.

The next proposition gives an expression of $\mathcal{S}$ to be computed.

Proposition 2.1. *For all* $v \in G_N$,

$$\mathcal{S}(v) = \prod_{i \in N} \bigcap_{\varepsilon \in R} [m_i(v_\varepsilon), M_i(v_\varepsilon)] = [q(v), Q(v)],$$

where for $i \in N$,

$$Q_i(v) \equiv M_i(v_{\varepsilon^i}) = \max\{\frac{v(\{i\}) + v(N) - v(N\setminus\{i\})}{2}, Q_i^1(v)\},$$

$$q_i(v) \equiv m_i(v_{\varepsilon^i}) = \min\{\frac{v(\{i\}) + v(N) - v(N\setminus\{i\})}{2}, q_i^1(v)\},$$

$$Q_i^1(v) \equiv \max_{i\in S, S\neq N,\{i\}} \{v(S) - v(S\setminus\{i\})\},$$

$$q_i^1(v) \equiv \min_{i\in S, S\neq N,\{i\}} \{v(S) - v(S\setminus\{i\})\},$$

and

$$\varepsilon^i \equiv \frac{v(\{i\}) - v(N) + v(N\setminus\{i\})}{2}.$$

Proof. Noting that by definition $Q_i^1(v_\varepsilon) = Q_i^1(v)$ and $q_i^1(v_\varepsilon) = q_i^1(v)$ for all $i \in N$ and all $\varepsilon \in R$,

$$\begin{aligned} M_i(v_\varepsilon) &= \max\{v(N) - v(N\setminus\{i\}) + \varepsilon, v(\{i\}) - \varepsilon, Q_i^1(v_\varepsilon)\} \\ &\geq \max\{\frac{v(\{i\}) + v(N) - v(N\setminus\{i\})}{2}, Q_i^1(v)\} \\ &= M_i(v_{\varepsilon^i}). \\ m_i(v_\varepsilon) &= \min\{v(N) - v(N\setminus\{i\}) + \varepsilon, v(\{i\}) - \varepsilon, q_i^1(v_\varepsilon)\} \\ &\leq \min\{\frac{v(\{i\}) + v(N) - v(N\setminus\{i\})}{2}, q_i^1(v)\} \\ &= m_i(v_{\varepsilon^i}). \end{aligned}$$

Hence for all $i \in N$ and all $\varepsilon \in R$,

$$\bigcap_{\varepsilon'\in R} [m_i(v_{\varepsilon'}), M_i(v_{\varepsilon'})] \subseteq [m_i(v_{\varepsilon^i}), M_i(v_{\varepsilon^i})] \subseteq [m_i(v_\varepsilon), M_i(v_\varepsilon)].$$

Hence for each $i \in N$,

$$\bigcap_{\varepsilon\in R} [m_i(v_\varepsilon), M_i(v_\varepsilon)] = [m_i(v_{\varepsilon^i}), M_i(v_{\varepsilon^i})]. \tag{2.1}$$

So we have the second equality:

$$\prod_{i\in N} \bigcap_{\varepsilon\in R} [m_i(v_\varepsilon), M_i(v_\varepsilon)] = \prod_{i\in N} [m_i(v_{\varepsilon^i}), M_i(v_{\varepsilon^i})] = [q(v), Q(v)].$$

We show the first equality For any fixed $\varepsilon \in R$,

$$\bigcap_{\varepsilon' \in R} [m_i(v_{\varepsilon'}), M_i(v_{\varepsilon'})] \subseteq [m_i(v_\varepsilon), M_i(v_\varepsilon)], \text{ for every } i \in N.$$

This implies

$$\prod_{i \in N} \bigcap_{\varepsilon' \in R} [m_i(v_{\varepsilon'}), M_i(v_{\varepsilon'})] \subseteq \prod_{i \in N} [m_i(v_\varepsilon), M_i(v_\varepsilon)] = \mathcal{R}(v_\varepsilon).$$

This implies

$$\prod_{i \in N} \bigcap_{\varepsilon' \in R} [m_i(v_{\varepsilon'}), M_i(v_{\varepsilon'})] \subseteq \bigcap_{\varepsilon \in R} \mathcal{R}(v_\varepsilon) = \mathcal{S}(v).$$

Next let $x \in \mathcal{S}(v)$. This implies $x \in \mathcal{R}(v_\varepsilon)$ for all $\varepsilon \in R$, which implies $x \in \mathcal{R}(v_{\varepsilon^i})$ for all $i \in N$. Hence for each $i \in N$,

$$m_i(v_{\varepsilon^i}) \leq \max_{j \in N}\{m_i(v_{\varepsilon^j})\} \leq x_i \leq \min_{j \in N}\{M_i(v_{\varepsilon^j})\} \leq M_i(v_{\varepsilon^i}).$$

From this and (2), we have

$$x_i \in [m_i(v_{\varepsilon^i}), M_i(v_{\varepsilon^i})] = \bigcap_{\varepsilon \in R} [m_i(v_\varepsilon), M_i(v_\varepsilon)],$$

which implies

$$x \in \prod_{i \in N} \bigcap_{\varepsilon \in R} [m_i(v_\varepsilon), M_i(v_\varepsilon)].$$

Hence

$$\mathcal{S}(v) \subseteq \prod_{i \in N} \bigcap_{\varepsilon \in R} [m_i(v_\varepsilon), M_i(v_\varepsilon)].$$

■

Gerard-Varet and Zamir [1] characterizes the solution $\mathcal{R}$ by some properties[5] . Here we characterize the solution $\mathcal{S}$.

Theorem 2.2. *A largest-under-inclusion solution satisfying* **SI** *and* **RE** *is the solution* $\mathcal{S}$.

[5] Gerard-Varet and Zamir [1] showed that $\mathcal{R}$ is a maximal reasonable correspondence.

Remark 2.1. *It is not difficult to prove that $\mathcal{PI}$ satisfies* **SI** *but does not satisfy* **RE**. *On the other hand, $\mathcal{R}$ satisfies* **RE** *but does not satisfy* **SI**. *Therefore, these two properties are independent.* *Proof.* Assume a solution ϕ satisfies **SI** and **RE**. By **RE** we have $\phi(v_\varepsilon) \subseteq \mathcal{R}(v_\varepsilon)$ for every $v \in G_N$ and every $\varepsilon \in R$. Hence by **SI**, we have $\phi(v) \subseteq \mathcal{R}(v_\varepsilon)$ since $\mathcal{R}(v_\varepsilon)$ is an interval in R^N. So $\phi(v) \subseteq \cap_{\varepsilon \in R}\mathcal{R}(v_\varepsilon) = \mathcal{S}(v)$. As we stated before, $\mathcal{S}$ satisfies **RE**. Furthermore, $Q_i(v_\varepsilon) = Q_i(v)$ and $q_i(v_\varepsilon) = q_i(v)$ for all $i \in N$, all $\varepsilon \in R$ and $v \in G_N$. These imply that $\mathcal{S}$ satisfies **SI**. ■

Proposition 2.3. *If a solution ϕ satisfies* **BDD**, **COV**, **MON** *and* **SI**, *and if $\phi(v) \neq \emptyset$ for all $v \in G_N$, then $\phi(v) \subseteq \mathcal{S}(v)$ for every $v \in G_N$.*

This corollary follows from Theorem 2.2 and Proposition 1.1.

3. Shift-Invariance and Various Solutions

In this section we see that very well-known solutions[6] are shift-invariant. We will see that solutions related to the pre-imputation set satisfy SI but solutions related to the imputation set do not. First we prove two lemmata which will be intensively used through this section. For a game $v \in G_N$, let[7]

$$\varepsilon_\sharp(v) \equiv -\frac{v(N) - \sum_{i \in N} v(\{i\})}{n}.$$

Lemma 3.1. *If a solution satisfies $\phi(v) \subseteq \mathcal{I}(v)$ for all $v \in G_N$ and for each $v \in G_N$ there exists $\varepsilon \in R$ such that $\phi(v_\varepsilon) \neq \emptyset$, then ϕ does not satisfy* **SI**.

Proof. For a game $v \in G_N$ and for $\varepsilon \in \mathbf{R}$, the set $\mathcal{I}(v_\varepsilon)$ is non-empty if and only if $\varepsilon \geq \varepsilon_\sharp(v)$. So, if $\varepsilon < \varepsilon_\sharp(v)$, then $\mathcal{I}(v_\varepsilon) = \emptyset$, which implies $\phi(v_\varepsilon) = \emptyset$ since $\phi(v_\varepsilon) \subseteq \mathcal{I}(v_\varepsilon)$. By the assumption there exists $\varepsilon' \in \mathbf{R}$ such that $\phi(v_{\varepsilon'}) \neq \emptyset$. So, $\ell_i(\phi(v_\varepsilon)) \neq \ell_i(\phi(v_{\varepsilon'}))$. This implies that ϕ does not satisfy **SI**. ■

---

[6] See Maschler [3], or Peleg and Sudhölter [7] to find the first papers where these solutions were defined.

[7] See p. 322 of Maschler, Peleg, and Shapley [4].

For any solution ϕ such that $\phi(v_\varepsilon) \neq \emptyset$ for some $\varepsilon \in R$, define two solutions $\overline{\phi}$ and $\underline{\phi}$ by

$$\overline{\phi}(v) \equiv \bigcup_{\varepsilon:\phi(v_\varepsilon)\neq\emptyset} \phi(v_\varepsilon) \text{ and } \underline{\phi}(v) \equiv \bigcap_{\varepsilon:\phi(v_\varepsilon)\neq\emptyset} \phi(v_\varepsilon), \forall v \in G_N.$$

Lemma 3.2. *For every solution ϕ such that $\phi(v_\varepsilon) \neq \emptyset$ for some $\varepsilon \in R$, the solutions $\overline{\phi}$ and $\underline{\phi}$ satisfy* **SI**.

Proof. By the definition of shifted game, we note that for any $v \in G_N$, $(v_\varepsilon)_\delta = v_{\varepsilon+\delta}$ for every $\varepsilon, \delta \in R$. For $v \in G_N$ and $\varepsilon \in R$,

$$\begin{aligned}\underline{\phi}(v) &= \bigcap_{\delta:\phi(v_\delta)\neq\emptyset} \phi(v_\delta) = \bigcap_{\delta:\phi(v_\delta)\neq\emptyset} \phi((v_\varepsilon)_{\delta-\varepsilon}) = \bigcap_{\lambda:\phi(v_{\lambda+\varepsilon})\neq\emptyset} \phi((v_\varepsilon)_\lambda) \\ &= \bigcap_{\lambda:\phi((v_\varepsilon)_\lambda)\neq\emptyset} \phi((v_\varepsilon)_\lambda) = \underline{\phi}(v_\varepsilon).\end{aligned}$$

In the same way we see $\overline{\phi}(v) = \overline{\phi}(v_\varepsilon)$. ■

Remark 3.1. *We assume that the empty set ϕ, that is, $\phi(v) = \emptyset$ for all $v \in G_N$, satisfies* **SI**.

Remark 3.2. *By the definition of $\mathcal{S}$ and Lemma 3.2 we see that $\mathcal{S} = \underline{\mathcal{R}}$.*

In the following we give examples of solutions which satisfy the shift-invariance.

3.1. The Shapley Value

The *Shapley value* is a solution which associates a vector $\varphi(v) \in R^N$ for $v \in G_N$ such that:

$$\begin{aligned}\varphi_i(v) &\equiv \sum_{S:i\in S} \frac{(|S|-1)!(n-|S|)!}{n!}[v(S) - v(S\backslash\{i\})] \\ &= \frac{v(N) - v(N\backslash\{i\}) + v(\{i\})}{n} + \sum_{S:i\in S, 2\leq|S|\leq n-1} \frac{(|S|-1)!(n-|S|)!}{n!}[v(S) - v(S\backslash\{i\})].\end{aligned}$$

It is well-known that the Shapley value satisfies **SI**. Since $\varphi_i(v)$ is a convex combination of $\{v(S) - v(S \setminus \{i\}) : \emptyset \neq S \subseteq N\}$, by definition we have $m_i(v) \leq \varphi_i(v) \leq M_i(v)$ for every $i \in N$. So $\varphi(v) \in \mathcal{R}(v)$. That is, the Shapley value satisfies **RE**. From Theorem 2.2, we have $\varphi(v) \in \mathcal{S}(v)$ for all $v \in G_N$.

3.2. The Least Core

The *core* of a game $v \in G_N$, denoted by $\mathcal{C}(v)$, is a well-known solution and it is defined by:

$$\mathcal{C}(v) = \{x \in R^N : x(N) = v(N) \text{ and } x(S) \geq v(S), \forall S \subset N\}.$$

By definition we see $\mathcal{C}(v) \subseteq \mathcal{I}(v)$ for all $v \in G_N$. For $\varepsilon \in R$, the *strong* ε *-core* of a game $v \in G_N$ is defined by the core of the ε-shifted game v_ε. The solution $\underline{\mathcal{C}}$ is called the *least-core*[8] . That is, the least core is the intersection of all nonempty strong ε-cores. Noting that for every $v \in G_N$, $\mathcal{C}(v_\varepsilon) \subseteq \mathcal{C}(v_{\varepsilon'})$ whenever $\varepsilon < \varepsilon'$. So there exists ε_0 such that $\mathcal{C}(v_\varepsilon) \neq \emptyset$ if and only if $\varepsilon \geq \varepsilon_0$. Then $\underline{\mathcal{C}}(v) = \mathcal{C}(v_{\varepsilon_0})$. So the least core is nonempty.

Proposition 3.1. *The least core satisfies* **SI** *and* $\underline{\mathcal{C}}(v) \subseteq \mathcal{S}(v)$ *for all* $v \in G_N$.

Proof. By Lemma 3.2, the least core satisfies **SI**. It is known that for $v \in G_N$, $x_i \leq M_i(v)$ for all $i \in N$ if $x \in \underline{\mathcal{C}}(v)$ [9] . Let's see that for $v \in G_N$, $x_i \geq m_i(v)$ for all $i \in N$ if $x \in \underline{\mathcal{C}}(v)$. Let $\underline{\mathcal{C}}(v) = \mathcal{C}(v_{\varepsilon_0})$. Assume $x \in \underline{\mathcal{C}}(v)$ and $x_i < m_i(v)$ for some $i \in N$. Then

$$\begin{aligned} &v(N \setminus \{i\}) - x(N \setminus \{i\}) < 0 < v(\{i\}) - x_i, \text{ and} \\ &\quad v(S \setminus \{i\}) - x(S \setminus \{i\}) < v(S) - x(S), \text{ for all } S : i \in S, n-1 \geq |S| \geq 2. \end{aligned}$$

Noting $\underline{\mathcal{C}}(v) = \mathcal{C}(v_{\varepsilon_0})$,

$$\varepsilon_0 = \max_{S \neq N, \emptyset} \{v(S) - x(S)\} = \max\{v(S) - x(S) : i \in S, 1 \leq |S| \leq n-1\}.$$

Define $y \in R^N$ by

$$y_j = \begin{cases} x_j - \delta, & \text{if } j \neq i; \\ x_i + (n-1)\delta, & \text{if } j = i, \end{cases}$$

where $\delta > 0$. Then

$$v(S) - y(S) = v(S) - x(S) - |N \setminus S|\delta < v(S) - x(S),$$

---

[8] See p. 306 of Maschler, Peleg and Shapley [4].

[9] See p. 182 of Peleg and Sudhölter [7]. Also at p.181-2 it is shown that each vector in the least core is individually rational for weakly superadditive (0-monotonic) games. The proof of it is applicable to the case here. But, for completeness, we give a proof.

for S such that $i \in S$ and $n-1 \geq |S| \geq 1$. If $i \notin S$ and $S \neq \emptyset$, then

$$v(S) - y(S) = v(S) - x(S) + |S|\delta < \max\{v(T) - x(T) : i \in T, T \neq N\}$$

by taking $\delta > 0$ which is sufficiently small. Hence $\max_{S \neq N, \emptyset}\{v(S) - y(S)\} < \varepsilon_0$. A contradiction. Hence $x_i \geq m_i(v)$ for all $i \in N$. Hence $\mathcal{C}(v) \subseteq [m(v), M(v)] = \mathcal{R}(v)$. From this and Theorem 2.2, we see that $\mathcal{C}(v) \subseteq \mathcal{S}(v)$ for all $v \in G_N$. ■

3.3. The Prekernel

For each $x \in R^N$ and $i, j \in N, i \neq j$, define the maximum surplus of i over j by $s_{ij}(x;v) \equiv \max\{v(S) - x(S) : i \in S, j \notin S\}$. The *prekernel* of $v \in G_N$ is a solution $\mathcal{PK}(v)$ defined by

$$\mathcal{PK}(v) \equiv \{x \in \mathcal{PI}(v) : s_{ij}(x;v) = s_{ji}(x;v), \quad \forall i, j \in N, i \neq j\}.$$

Proposition 3.2. *The prekernel $\mathcal{PK}$ satisfies* **SI** *and $\mathcal{PK}(v) \subseteq \mathcal{S}(v)$ for all $v \in G_N$.*

Proof. It is known that[10] $\mathcal{PK}(v) = \mathcal{PK}(v_\varepsilon)$ for every $v \in G_N$ and every $\varepsilon \in R$. This means that the prekernel satisfies **SI**. Since the prekernel satisfies[11] **RE**, by Proposition 2.1, we see that $\mathcal{PK}(v) \subseteq \mathcal{S}(v)$ for all $v \in G_N$. ■

Remark 3.3. *The kernel of $v \in G_N$ is a solution $\mathcal{K}(v)$ defined by*

$$\mathcal{K}(v) \equiv \{x \in \mathcal{I}(v) : \textit{For every } i, j \in N, i \neq j, s_{ij}(x;v) \leq s_{ji}(x;v) \textit{ or } x_j = v(\{j\})\}.$$

The kernel satisfies **RE**, *but it does not satisfy* **SI** *by Lemma 3.1.*

3.4. The Prenucleolus

Let $v \in G_N$ be a game. For $x \in R^N$ let $\theta(x)$ be the 2^n-vector whose components are the numbers $v(S) - x(S), S \subseteq N$, arranged in nonincreasing order, i.e., $\theta_i(x) \geq \theta_j(x)$ whenever $1 \leq i \leq j \leq 2^n$. We say that $\theta(x)$ is lexicographically smaller than $\theta(y)$, denoted $\theta(x) \prec_L \theta(y)$, if and only if there is an index k such that

---

[10] See p. 74 of Pechersky and Sobolev [6].

[11] See p. 131 of Peleg and Sudhölter [7].

$\theta_i(x) = \theta_i(y)$ for all $i < k$, and $\theta_k(x) < \theta_k(y)$. We write $\theta(x) \preceq_L \theta(y)$ for not $\theta(y) \prec_L \theta(x;)$. The *prenucleolus* of v is defined by

$$\mathcal{PN}(v) = \{x \in \mathcal{PI}(v) : \theta(x) \preceq_L \theta(y) \text{ for all } y \in \mathcal{PI}(v)\}.$$

The *nucleolus* of v is defined by

$$\mathcal{N}(v) = \{x \in \mathcal{I}(v) : \theta(x) \preceq_L \theta(y) \text{ for all } y \in \mathcal{I}(v)\}.$$

It is known[12] that $\mathcal{PN}(v)$ and $\mathcal{N}(v)$ consist of one vector respectively and $\mathcal{N}(v) \subseteq \mathcal{K}(v)$ and $\mathcal{PN}(v) \subseteq \mathcal{PK}(v)$. By definition it holds $\mathcal{PN}(v) = \mathcal{N}(v)$ when $\mathcal{PN}(v) \subseteq \mathcal{I}(v)$.

Proposition 3.3.[13] *The prenucleolus $\mathcal{PN}$ satisfies* **SI** *and $\mathcal{PN}(v) \subseteq \mathcal{S}(v)$ for all $v \in G_N$.*

Proof. It is known that $\mathcal{PN}(v) = \mathcal{PN}(v_\varepsilon)$ for every $v \in G_N$ and every $\varepsilon \in R$. This means that the prenucleolus satisfies **SI**. Since the prenucleolus is included in the prekernel, we have $\mathcal{PN}(v) \subseteq \mathcal{S}(v)$ for all $v \in G_N$ by Proposition 3.3. ■

Remark 3.4. *For a game $v \in G_N$, since $\mathcal{I}(v_\varepsilon) \subseteq \mathcal{I}(v_\delta)$ for any $\varepsilon < \delta$, there exists $\varepsilon_3 \in R$ such that $\mathcal{PN}(v) \in \mathcal{I}(v_\varepsilon)$ if and only if $\varepsilon \geq \varepsilon_3$. So $\mathcal{PN}(v) = \mathcal{N}(v_\varepsilon)$ for $\varepsilon \geq \varepsilon_3$. Note that $\varepsilon_3 \geq \varepsilon_\sharp$ because $\mathcal{I}(v_{\varepsilon_3}) \neq \emptyset$. Furthermore, $\mathcal{N}(v_{\varepsilon_\sharp}) = \mathcal{I}(v_{\varepsilon_\sharp}) = \mathcal{E}(v)$ and $\mathcal{N}(v_\varepsilon) = \emptyset$ for $\varepsilon < \varepsilon_\sharp$. Hence $\overline{\mathcal{N}(v)}$ is a finite curve whose extreme points are $\mathcal{PN}(v)$ and $\mathcal{E}(v)$. By Lemma 3.2, $\overline{\mathcal{N}}$ satisfies* **SI***.*

4. Pairwise Reasonable Outcomes

Finally we see another solution which satisfies **SI**. A player $i \in N$ is said to be *at least as desirable as* a player $j \in N$ in a game $v \in G_N$ if $v(S \cup \{i\}) \geq v(S \cup \{j\})$ for all $S \subseteq N \setminus \{i, j\}$.

Preservation of desirability (**PD**): For all $v \in G_N$, all $i, j \in N$, whenever i is at least as desirable as j, it holds $x_i \geq x_j$ for every $x \in \phi(v)$.

12 See Maschler [3], p. 612.

13 Also, see Remark 9 of Pechersky and Sobolev [6].

A vector $x \in R^N$ is said to be *pairwise reasonable*[14] if for every pair of players $i, j \in N$, the payoff x_i for player i should not exceed the payoff x_j for player j by more than the greatest amount that the contribution of player i to any coalition exceeds the contribution of player j to that same coalition, or

$$x_i - x_j \leq \bar{t}_{ij}(v) \equiv \max_{S \subseteq N \setminus \{i,j\}} [v(S \cup \{i\}) - v(S \cup \{j\})],$$

or equivalently

$$x_j - x_i \geq \underline{t}_{ji}(v) \equiv \min_{S \subseteq N \setminus \{i,j\}} [v(S \cup \{j\}) - v(S \cup \{i\})].$$

For a game $v \in G_N$ we denote by $\mathcal{PR}(v)$ the set of all pairwise reasonable vectors. $\mathcal{PR}$ is a solution.

Proposition 4.1. *The solution $\mathcal{PR}$ is a largest-under-inclusion solution satisfying* **COV** *and* **PD**. *It satisfies also* **SI**.

Remark 4.1. *We see that three properties in Proposition 4.1 are independent by using solutions defined in the remark which is located above Proposition 1.1. The solution ϕ^1 satisfies* **COV** *and* **SI**, *and it does not satisfy* **PD**. *The solution ϕ^2 satisfies* **PD** *and* **SI**, *and it does not satisfy* **COV**. *The solution ϕ^3 satisfies* **PD** *and* **COV**, *and it does not satisfy* **SI**. *So* **COV**, **PD** *and* **SI** *are independent.*

This proposition follows from the next two lemmas.

Lemma 4.1. *Assume a solution ϕ satisfies* **COV** *and* **PD**. *It holds $\phi(v) \subseteq \mathcal{PR}(v)$ for every $v \in G_N$.*

Proof. Assume a solution ϕ satisfies **COV** and **PD**. For a game $v \in G_N$, if $\phi(v) = \emptyset$ then trivially $\phi(v) \subseteq \mathcal{PR}(v)$. Assume $\phi(v) \neq \emptyset$ and fix $y \in \phi(v)$. Fix any $i, j \in N$. Let

$$w(S) = \begin{cases} v(S) - \underline{t}_{ij}(v), & \text{if } i \in S \\ v(S), & \text{if } i \notin S. \end{cases}$$

Then $w(S \cup \{i\}) - w(S \cup \{j\}) = v(S \cup \{i\}) - v(S \cup \{j\}) - \underline{t}_{ij}(v) \geq 0$ for every $S \subseteq N \setminus \{i, j\}$. By **PD**, $x_i \geq x_j$ for every $x \in \phi(w)$. By **COV**, $\phi(w) = \phi(v) - \underline{t}_{ij}(v)e_i$. Let $x = y - \underline{t}_{ij}(v)e_i$. Then $x \in \phi(w)$ and $y_i = x_i + \underline{t}_{ij}(v) \geq x_j + \underline{t}_{ij}(v) = y_j + \underline{t}_{ij}(v)$

[14] See pp. 606-607 of Maschler [3].

which implies $y_i - y_j \geq \underline{t}_{ij}(v)$. Consequently we see $y_i - y_j \geq \underline{t}_{ij}(v)$ for every fixed $i, j \in N$. Hence $y \in \mathcal{PR}(v)$ and so $\phi(v) \subseteq \mathcal{PR}(v)$. ■

Lemma 4.2. *A solution* $\mathcal{PR}$ *satisfies* **COV**, **PD** *and* **SI**.

Proof. Let $v \in G_N$.

COV: For $k > 0$ and $y \in R^N$, it is easy to see that $\underline{t}_{ij}(kv+y) = k\underline{t}_{ij}(v) + y_i - y_j$. So we have that $x \in \mathcal{PR}(v)$ if and only if $kx + y \in \mathcal{PR}(kv+y)$. Hence $\mathcal{PR}(v) + y = \mathcal{PR}(kv+y)$.

PD: For any $i, j \in N$, if $v(S \cup \{i\}) \geq v(S \cup \{j\})$ for every $S \subseteq N \setminus \{i, j\}$, then $\underline{t}_{ij}(v) \geq 0$. So $x_i - x_j \geq \underline{t}_{ij}(v) \geq 0$ if $x \in \mathcal{PR}(v)$.

SI: This is from the fact that for any $i, j \in N$ and $\varepsilon \in R$, $v(S \cup \{i\}) - v(S \cup \{j\}) = v_\varepsilon(S \cup \{i\}) - v_\varepsilon(S \cup \{j\})$ for every $S \subseteq N \setminus \{i, j\}$. ■

Remark 4.2. *The solution* $\mathcal{PR}$ *does not satisfy* **RE**. *For example, let* $n = 3$ *and* $v(N) = v(23) = 3, v(12) = v(13) = \frac{1}{2}$ *and* $v(1) = v(2) = v(3) = 0$. *Then a vector* $(1, 1, 1)$ *is in* $\mathcal{PR}(v)$ *and it is not in* $\mathcal{R}(v)$.

It is known that the Shapley value and the kernel are in the set of all pairwise reasonable vectors[15]. Since the prekernel satisfies **PD**[16] and **COV**, it is in the set of all pairwise reasonable vectors. A 3-person game given in the above remark is an example where the least core is not in the set of all pairwise reasonable vectors.

References

[1] L.A. Gerard-Valet and S. Zamir, Remarks on the reasonable set of outcomes in a general coalition function game, *International Journal of Game Theory* **16** (1987), 123–143.

[2] F. Llerena and C. Rafels, *On reasonable outcomes and the core*, 2006, mimeo.

[3] M. Maschler, The bargaining set, kernel, and nucleolus, *Handbook of Game Theory* **1** (1992), Edited by R.J.Aumann and S.Hart.

[4] M. Maschler, B. Peleg and L.S. Shapley, Geometric properties of the kernel, nucleolus, and related solution concepts, *Mathematics of Operations Research* **4** (1979), 303–338.

[15] See p. 607 of Maschler [3].

[16] See p. 120 of Peleg and Sudhölter [7].

[5] J.W. Milnor, *Reasonable outcomes for n-person games*, RM-916, The Rand Corporation, Santa Monica, CA, 1952.

[6] S. Pechersky and A. Sobolev, Set-valued nonlinear analogues of the Shapley value, *International Journal of Game Theory* **24** (1995), 57–78.

[7] B. Peleg and P. Sudhölter, *Introduction to the Theory of Cooperative Games*, Second Edition, Kluwer Academic Publishers, Boston, MA, 2007.

[8] L.S. Shapley and M. Shubik, Quasi-cores in a monetary economy with nonconvex preferences, *Econometrica* **34** (1966), 805–827.

[9] S.H. Tijs Bounds for the core and the τ-value, In :Moeschlin O, Pallaschke D (eds), *Game Theory and Mathematical Economic,* Noth Holland Publishing Company, Amsterdam, 1981.

In: Mathematical Theories, Models and Groups ISBN 978-1-61324-250-6
Editor: Marcus J. Grecho

Chapter 6

GROUPS OF WRAPS AND DIFFEOMORPHISMS OF MANIFOLDS OVER NON-ARCHIMEDEAN NORMED FIELDS

S. V. Ludkovsky*
Department of Applied Mathematics,
Moscow State Technical University MIREA,
Moscow, Russia

Abstract

The article is devoted to the investigation of groups of diffeomorphisms and wraps of manifolds over non-archimedean fields of zero and positive characteristics. Different types of topologies are considered on groups of wraps and diffeomorphisms relative to which they are generalized Lie groups or topological groups. Among such topologies pairwise incomparable are found as well. Topological perfectness of the diffeomorphism group relative to certain topologies is studied. There are proved theorems about projective limit decompositions of these groups and their compactifications for compact manifolds. Moreover, an existence of one-parameter local subgroups of diffeomorphism groups is investigated.

*E-mail address: sludkowski@mail.ru

1. Introduction

Non-archimedean analysis has rather long history, but it is much less developed in comparison with the classical analysis over the fields **R** and **C**. Therefore, the theory of groups on manifolds over non-archimedean fields is not so well investigated as for Riemann or complex manifolds [35, 16, 12, 5].

As it is known fields with multiplicative ultra-norms such as the field of p-adic numbers were first introduced by K. Hensel [14]. Several years later on it was proved by A. Ostrowski [33] that on the field of rational numbers each multiplicative norm is either the usual norm as in **R** or is equivalent to a non archimedean norm $|x| = p^{-k}$, where $x = np^k/m \in \mathbf{Q}$, $n, m, k \in \mathbf{Z}$, $p \geq 2$ is a prime number, n and m and p are mutually pairwise prime numbers. Each locally compact infinite field with a non trivial non archimedean valuation is either a finite algebraic extension of the field of p-adic numbers or is isomorphic to the field $\mathbf{F}_{p^k}(\theta)$ of power series of the variable θ with expansion coefficients in the finite field $\mathbf{F}_{p^k}$ of p^k elements, where $p \geq 2$ is a prime number, $k \in \mathbf{N}$ is a natural number [38, 43]. The valuation group $\Gamma_{\mathbf{K}} := \{|x| : x \in \mathbf{K}, x \neq 0\} \subset (0, \infty)$ of a locally compact field **K** is discrete. Non locally compact fields are wide spread as well and among them there are fields with the valuation group $\Gamma_{\mathbf{K}} = (0, \infty)$ [9, 38, 40]. The non-archimedean analog of the field of complex numbers is $\mathbf{C_p}$ which is complete algebraically and as the uniform space relative to its multiplicative norm and $\Gamma_{\mathbf{C_p}} = \{x \in \mathbf{Q} : x > 0\}$.

The importance of transformation groups of manifolds in the non-Archimedean functional analysis, representation theory and mathematical physics is clear and also can be found in the references given below [2, 21, 22, 24, 25, 38, 39]. Here we use the terminology a wrap group instead of a (geometric) loop group, because in the non-archimedean case there are not any circles like as in classical case, moreover, manifolds are already totally disconnected and of arbitrary dimension over a non-archimedean field. It was shown earlier in preceding works of the author that such groups are smooth manifolds with differentiable group operations depending on a considered class of smoothness, that is they are Lie groups.

This article is devoted to several aspect of such groups and to further investigation of their structure. One of them is on their structure from the point of view of the non-archimedean compactficiation (see also about Banaschewski compactification in [38]). Though a new topology used for compactification may be different from the initial topology of a group or even may be non-comparable, because on the same group it is possible an existence of several different topologies making it a topological group. This is useful also for studying their representations as restrictions of representations of non-archimedean compactifications, which are constructed below such that they also are groups. Apart from previous works [21, 22, 23, 24, 25], where the characteristic $char(\mathbf{K}) = 0$ was zero, in this paper groups on manifolds over fields with non-zero characteristics also are defined and investigated. Different types of topologies are considered on groups of wraps and diffeomorphisms relative to which they are generalized Lie groups or topological groups. Among such topologies pairwise incomparable are found as well. It is caused by the fact that repeated application of projective and inductive limits of topological spaces generate topologies and spaces in general dependent from an order of taking limits and their types, so that such topologies may appear incomparable on a subset contained in these topological spaces. It is proved, that relative to the C^∞ bounded-open topology groups of geometric wraps and groups of diffeomorphisms of manifolds over non-archimedean ultra-normed fields are the generalized Lie groups.

Previously one-parameter subgroups over fields of non-zero characteristics were not studied. This article contains as well results on one-parameter subgroups over $\mathbf{K}$ with $char(\mathbf{K}) = p > 1$ using its multiplicative subgroup $\mathbf{K}^* := \mathbf{K} \setminus \{0\}$. It is proved below that the diffeomorphism group of a compact manifold is topologically simple relative to the C^∞ compact-open topology, that develops previous results [24], where topological simplicity and perfectness was proved over fields of zero characteristic.

At first in Section 2 we remind basic facts and notations, which are given in detail in references [38, 40, 21, 22, 23]. A wrap group $L_t(M,N)$ is defined as a quotient space of a family of mappings $f : M \to N$ of class C^t of one Banach manifold M into another N over the same local field $\mathbf{K}$ such that $\lim_{z\to s}(\bar{\Phi}^m f)(z;h_1,...,h_m;\zeta_1,...,\zeta_m) = 0$ or $\lim_{z\to s}(\Upsilon^m f)(z^{[m]}) = 0$ for each $0 \le m \le t$, where M and N are embedded into the corresponding Banach spaces X and Y, $cl(M) = M \cup \{s\}$, $cl(M)$ and N are clopen in X and Y respectively, $0 \in N$, $(\bar{\Phi}^m f)(z;h_1,...,h_m;\zeta_1,..,\zeta_m)$ and $(\Upsilon^m f)(z^{[m]})$ are continuous extensions of difference quotients by variables corresponding to z or by all appearing inductively variables over a non-archimedean field $\mathbf{K}$ of zero characteristic or of $char(\mathbf{K}) = p > 0$ respectively, $z \in M$, $h_1,...,h_m$ are nonzero vectors in X, $\zeta_1,...,\zeta_m \in \mathbf{K}$ such that $z+\zeta_1 h_1+...+\zeta_m h_m \in M$, $z^{[m+1]} := (z^{[m]}, v^{[m]}, \zeta_{m+1})$, $z^1 = (z, v^{[0]}, \zeta_1)$, $z^{[m]}, v^{[m]} \in \mathbf{K}^{[m]}$, $\mathbf{K}^{[m+1]} = \mathbf{K}^{[m]} \oplus \mathbf{K}^{[m]} \oplus \mathbf{K}$, $U^{[1]} = U$, $z^{[m]} + \zeta_{m+1} v^{[m]} \in U^{[m]}$ (see also §2.1).

In Section 2 preliminary investigations on structures of $Diff^t(M)$ as the topological groups and Lie groups are studied. Non-archimedean completions of clopen subgroups W of wrap groups G and diffeomorphism groups G are considered in Sections 3 and 7. Completions are considered relative to uniformities associated with projective decompositions. They produce topologies incomparable with the initial one. Relative to them they remain topological groups. In the case of the wrap group the non-archimedean completion produces a new topological group V in which the initial group W is embedded as a dense subgroup such that $V \neq W$. Such topologies have purely non-archimedean origin related with non-archimedean uniformities or families of non-archimedean semi-norms on spaces of continuous or more narrow classes of functions between non-archimedean manifolds. In the classical case over **R** one might expect instead of this some repeated combination of an inductive and a projective limits, which is quite different thing.

For the compact manifold M in the case of the diffeomorphism group the non-archimedean completion of W produces profinite group. For the locally compact manifolds M and N in the case of the wrap group $L_t(M,N)$ one of the non-archimedean completion of W produces its embedding into $\mathbf{Z_p}^{\mathbf{N}}$ and also there exists the completion isomorphic with $(\nu\mathbf{Z})^{\aleph_0}$, where $\nu\mathbf{Z}$ is the one-point Alexandroff compactification of **Z**. When W is bounded relative to the corresponding metric in $L_t(M,N)$, then W is embedded into $\mathbf{Z_p}^{\mathbf{N}}$. Moreover, topologies of $Diff^w(M)$ and $Diff^t(M)$ or $L_t(M,N)$ and $L^w(M,N)$ are incomparable for compact manifolds M and N, where the groups $Diff^w(M)$ and $L^w(M,N)$ are supplied with the weak projective limit topologies τ_w. The group $Diff^t(M)$ is topologically simple, on the other hand, the group $L_t(M,N)$ is commutative.

An existence of one parameter subgroups of $Diff^t(M)$ is investigated in Section 5. It is proved in Section 6, that $Diff^t(M)$ is topologically simple relative to its C^t compact-open topology, as well as the theorem about continuous automorphisms of $Diff^t(M)$ is proved.

The notation given below and the corresponding definitions are given in detail in [21, 23]. All results of this paper over the fields of positive characteristics are obtained for the first time.

2. Groups of Diffeomorphisms

1. Definitions. Let **K** be an infinite field with a non trivial non archimedean valuation, let also X and Y be topological vector spaces over **K** and U be an open subset in X. For a function $f : U \to Y$ consider the associated function

$f^{[1]}(x,v,t) := [f(x+tv) - f(x)]/t$

on a set $U^{[1]}$ at first for $t \neq 0$ such that $U^{[1]} := \{(x,v,t) \in X^2 \times \mathbf{K}, x \in U, x + tv \in U\}$. If f is continuous on U and $f^{[1]}$ has a continuous extension on $U^{[1]}$, then we say, that f is continuously differentiable or belongs to the class C^1. The $\mathbf{K}$-linear space of all such continuously differentiable functions f on U is denoted $C^{[1]}(U,Y)$. By induction we define functions $f^{[n+1]} := (f^{[n]})^{[1]}$ and spaces $C^{[n+1]}(U,Y)$ for $n = 1,2,3,...$, where $f^{[0]} := f$, $f^{[n+1]} \in C^{[n+1]}(U,Y)$ has as the domain $U^{[n+1]} := (U^{[n]})^{[1]}$.

The differential $df(x) : X \to Y$ is defined as $df(x)v := f^{[1]}(x,v,0)$.

Define also partial difference quotient operators Φ^n by variables corresponding to x only such that

$\Phi^1 f(x;v;t) = f^{[1]}(x,v,t)$

at first for $t \neq 0$ and if $\Phi^1 f$ is continuous for $t \neq 0$ and has a continuous extension on $U^{[1]} =: U^{(1)}$, then we denote it by $\bar{\Phi}^1 f(x;v;t)$. Define by induction

$\Phi^{n+1} f(x;v_1,...,v_{n+1};t_1,...,t_{n+1}) := \Phi^1(\Phi^n f(x;v_1,...,v_n;t_1,...,t_n))(x;v_{n+1};t_{n+1})$

at first for $t_1 \neq 0,...,t_{n+1} \neq 0$ on $U^{(n+1)} := \{(x;v_1,...,v_{n+1};t_1,...,t_{n+1}) : x \in U; v_1,...,v_{n+1} \in X; t_1 ...,t_{n+1} \in \mathbf{K}; x+v_1t_1 \in U,...,x+v_1t_1+...+v_{n+1}t_{n+1} \in U\}$. If f is continuous on U and partial difference quotients $\Phi^1 f$,...,$\Phi^{n+1} f$ has continuous extensions denoted by $\bar{\Phi}^1 f$,..., $\bar{\Phi}^{n+1} f$ on $U^{(1)}$,...,$U^{(n+1)}$ respectively, then we say that f is of class of smoothness C^{n+1}. The $\mathbf{K}$ linear space of all C^{n+1} functions on U is denoted by $C^{n+1}(U,Y)$, where $\Phi^0 f := f$, $C^0(U,Y)$ is the space of all continuous functions $f : U \to Y$.

Then the differential is given by the equation $d^n f(x).(v_1,...,v_n) := n!\bar{\Phi}^n f(x;v_1,...,v_n;0,...,0)$, where $n \geq 1$, also denote $D^n f = d^n f$. Shortly we shall write the argument of $f^{[n]}$ as $x^{[n]} \in U^{[n]}$ and of $\bar{\Phi}^n f$ as $x^{(n)} \in U^{(n)}$, where $x^{[0]} = x^{(0)} = x$, $x^{[1]} = x^{(1)} = (x,v,t)$, $v^{[0]} = v^{(0)} = v$, $t_1 = t$, $x^{[k]} = (x^{[k-1]}, v^{[k-1]}, t_k)$ for each $k \geq 1$, $x^{(k)} := (x;v_1,...,v_k;t_1,...,t_k)$.

Subspaces of uniformly C^n or $C^{[n]}$ bounded continuous functions together with $\bar{\Phi}^k f$ or $\Upsilon^k f$ on bounded open subsets of U and $U^{(k)}$ or $U^{[k]}$ for $k = 1,...,n$ we denote by $C^n_b(U,Y)$ or $C^{[n]}_b(U,Y)$ respectively.

Consider partial difference quotients of products and compositions of functions and relations between partial difference quotients and differentiability of both types. Denote by $L(X,Y)$ the space of all continuous $\mathbf{K}$-linear mappings $A : X \to Y$. By $L_n(X^{\otimes n},Y)$ denote the space of all continuous $\mathbf{K}$ n-linear mappings $A : X^{\otimes n} \to Y$, particularly, $L(X,Y) = L_1(X^{\otimes 1},Y)$. If X and Y are normed spaces, then $L_n(X^{\otimes n},Y)$ is supplied with the operator norm:

$\|A\| := \sup_{h_1\neq 0,\dots,h_n\neq 0; h_1,\dots,h_n\in X} \|A.(h_1,\dots,h_n)\|_Y/(\|h_1\|_X \dots \|h_n\|_X)$.

2. Lemma. *The spaces $C^{[1]}(U,Y)$ and $C^1(U,Y)$ are linearly topologically isomorphic. If $f \in C^n(U,Y)$, then $\bar{\Phi}^n f(x;*;0,\dots,0) : X^{\otimes n} \to Y$ is a* **K** *n-linear $C^0(U,L_n(X^{\otimes n},Y))$ symmetric map.*

Proof. From Definition 1 it follows, that $f^{[1]}(x,v,t) = \bar{\Phi}^1 f(x;v;t)$ on $U^{[1]} = U^{(1)}$, so both **K**-linear spaces are linearly topologically isomorphic. On the other hand, due to its definition $\bar{\Phi}^n f(x;*,0,\dots,0)$ is the **K** n-linear symmetric mapping for each $x \in U$ and it belongs to $C^0(U,L_n(X^{\otimes n},Y))$, since $\bar{\Phi}^n f(x;v_1,\dots,v_n;t_1,\dots,t_n)$ is continuous on $U^{(n)}$ and for each $x \in U$ and $v_1,\dots,v_n \in X$ there exist neighborhoods V_i of v_i in X and W of zero in **K** such that $x + WV_1 + \dots + WV_n \subset U$.

3. Lemma. *Operators $\Upsilon^n(f) := f^{[n]}$ from $C^{[n]}(U,Y)$ into $C^0(U^{[n]},Y)$ and $\bar{\Phi}^n : C^n(U,Y) \to C^0(U^{(n)},Y)$ are* **K***-linear and continuous.*

Proof. Since $[(af+bg)(x+vt)-(af+bg)(x)]/t = a(f(x+vt)-f(x))/t + b(g(x+vt)-g(x))/t$ for each $f,g \in C^1(U,Y)$ and each $a,b \in$ **K**, then applying this formula by induction and using definitions of operators Υ^n and $\bar{\Phi}^n$ we get their **K**-linearity. Indeed,

$\Upsilon^n(af+bg)(x^{[n]}) = \Upsilon^1(\Upsilon^{n-1}(af+bg)(x^{[n-1]}))(x^{[n]}) = \Upsilon^1(af^{[n-1]} + bg^{[n-1]})(x^{[n]}) = af^{[n]}(x^{[n]}) + bg^{[n]}(x^{[n]})$ and

$\bar{\Phi}^n(af+bg)(x^{(n)}) = \bar{\Phi}^1(\bar{\Phi}^{n-1}(af+bg)(x^{(n-1)}))(x^{(n)}) = \bar{\Phi}^1(af^{(n-1)} + bg^{(n-1)})(x^{(n)}) = af^{(n)}(x^{(n)}) + bg^{(n)}(x^{(n)})$.

The continuity of Υ^n and $\bar{\Phi}^n$ follows from definitions of spaces $C^{[n]}(U,Y)$ and $C^n(U,Y)$ respectively.

4. Definitions. Let M be a manifold modelled on a topological vector space X over **K** such that its atlas $At(M) := \{(U_j, {}_M\phi_j) : j \in \Lambda_M\}$ is of class $C^{\alpha'}_{\beta'}$, that is the following four conditions are satisfied:

$(M1)$ $\{U_j : j \in \Lambda_M\}$ is an open covering of M, $U_j = {}_MU_j$,

$(M2)$ $\bigcup_{j\in\Lambda_M} U_j = M$,

$(M3)$ ${}_M\phi_j := \phi_j : U_j \to \phi_j(U_j)$ is a homeomorphism for each $j \in \Lambda_M$, $\phi_j(U_j) \subset X$,

$(M4)$ $\phi_j \circ \phi_i^{-1} \in C^{\alpha'}_{\beta'}$ on its domain for each $U_i \cap U_j \neq \emptyset$,

where Λ_M is a set, $C^\infty := \bigcap_{l=1}^\infty C^l_\beta$, $C^{[\infty]}_\beta := \bigcap_{l=1}^\infty C^{[l]}$, $\alpha' \in \{n,[n] : 1 \le n \le \infty\}$, $\beta \in \{\emptyset, b\}$, $C^{\alpha'}_\emptyset := C^{\alpha'}$.

Supply $C^\alpha_\beta(U,Y)$ with the bounded-open C^α_β topology (denoted by $\tau_{\alpha,\beta}$ generally or τ_α for $\beta=\emptyset$ or for compact U) with the base $W(P,V)=\{f\in C^\alpha_\beta(X,Y): S^k f|_P\in V, k=0,...,n\}$ of neighborhoods of zero, where P is bounded and open in $U\subset X$, $P\subset U$, V is open in Y, $0\in V$, $S^k=\bar{\Phi}^k$ or $S^k=\Upsilon^k$ for $\alpha=n$ or $\alpha=[n]$ respectively, $v_1,...,v_n\in(P-y_0)$, $v_l^{[k]}\in(P-y_0)$ for each k,l for some marked $y_0\in P$ and $|t_j|\leq 1$ for every j.

If M and N are $C^{\alpha'}_\beta$ manifolds on topological vector spaces X and Y over $\mathbf{K}$, then consider the uniform space $C^\alpha_\beta(M,N)$ of all mappings $f:M\to N$ such that $f_{j,i}\in C^\alpha_\beta$ on its domain for each $j\in\Lambda_N$, $i\in\Lambda_M$, where $f_{j,i}:={}_N\phi_j\circ f\circ {}_M\phi_i^{-1}$ is with values in Y, $\alpha\leq\alpha'$. The uniformity in $C^\alpha_\beta(M,N)$ is inherited from the uniformity in $C^\alpha_\beta(X,Y)$ with the help of charts of atlases of M and N. If M is compact, then $C^\alpha_b(M,N)$ and $C^\alpha(M,N)$ coincide.

The family of all homeomorphisms $f:M\to M$ of class C^α_β denote by $Diff^\alpha_\beta(M)$.

Let γ be a set, then denote by $c_0(\gamma,\mathbf{K})$ the normed space consisting of all vectors $x=\{x_j\in\mathbf{K}: j\in\gamma$, for each $\varepsilon>0$ the set $\{j:|x_j|>\varepsilon\}$ is finite $\}$, where $\|x\|:=\sup_{j\in\gamma} x_j|$. In view of the Kuratowski-Zorn lemma it is convenient to consider γ as an ordinal. Henceforth, suppose that $X=c_0(\gamma_X,\mathbf{K})$ and $Y=c_0(\gamma_Y,\mathbf{K})$.

5. Theorem. *The uniform space $Diff^\alpha_b(M)$ (see §4) is the topological group relative to compositions of mappings.*

Proof. The group operation in $Diff^\alpha_\beta(M)$ is $(f,g)\mapsto f\circ g$, where $f\circ g(x):=f(g(x))$ for each $x\in M$. Then $f=id$ is the unit element in $Diff^\alpha_\beta(M)$, where $id(x)=x$ for each $x\in M$. Since the composition of mappings is associative, then $f\circ(g\circ h)=(f\circ g)\circ h$ is associative as the group operation. For each $f\in Diff^\alpha_\beta$ there exists its inverse mapping f^{-1} such that $f^{-1}(y)=x$ for each $y=f(x)$, $x\in M$, since $f:M\to M$ is the homeomorphism. It remains to verify that $f^{-1}\in Diff^\alpha_\beta(M)$ for each $f\in Diff^\alpha_\beta(M)$ and the composition $(Diff^\alpha_\beta)^2\ni(f,g)\mapsto f\circ g\in Diff^\alpha_\beta(M)$ and inversion $f\mapsto f^{-1}$ are continuous operations.

In the normed space $Y=c_0(\gamma_Y,\mathbf{K})$ a subspace $span_{\mathbf{K}}\{e_j: j\in\gamma_X\}$ consisting of all finite $\mathbf{K}$-linear combinations of vectors $e_j=(0,...,0,1,0,...)$ with 1 on the j-th place is everywhere dense. Therefore, each $f\in C^\alpha_b(U,Y)$ is the uniform limit of mappings $(f_1,...,f_j,0,...)\in C^\alpha_b(U,Y)$ together with $\Phi^k f$ or $\Upsilon^k f$ on bounded subsets of U and $U^{(k)}$ or $U^{[k]}$ for $1\leq k\leq n$, $n\in\mathbf{N}$, $n\leq\alpha$. In particular, consider ${}_N\phi_l\circ f\circ {}_M\phi_i^{-1}$ for $f\in C^\alpha_b(M,N)$ taking U as a finite union of ${}_M\phi_i({}_MU_j)$. Consider all possible embeddings of $\mathbf{K}^v$ into X, particularly, containing $x^{(k)}$ or $x^{[k]}$ for each $0 \le k \le n$, where $n \in \mathbf{N}$, $n \le \alpha$. In view of Formulas 6(1) or 7(1) of the Appendix using restrictions on different embedded subspaces $\mathbf{K}^{[k]}$ or $\mathbf{K}^{(k)}$ into $X^{[k]}$ or $X^{(k)}$ and uniform continuity of Υ^k and $\bar{\Phi}^k$ on bounded open subsets, $k = 0,1,2,...$, we get, that $f \circ g \in C^\alpha_b(M,M)$ for each $f,g \in Diff^\alpha_b(M)$, since $f_{l,s} \circ g_{s,i} \in C^\alpha_b(U_{l,s,i},X)$ on corresponding domains $U_{l,s,i}$ in X. From $f,g \in Hom(M,M)$ it follows, that $f \circ g \in Hom(M,M)$, hence $f \circ g \in Diff^\alpha_b(M)$. Applying to $id_{l,i} = f^{-1}_{l,s} \circ f_{s,i}$ on corresponding domains Formulas A.6(1) or A.7(1) and restricting on different embedded subspaces $\mathbf{K}^{[k]}$ or $\mathbf{K}^{(k)}$ in $X^{[k]}$ or $X^{(k)}$ and using uniform continuity on bounded open subsets for Υ^k or $\bar{\Phi}^k$, $k = 0,1,2,...$, to both sides of this equality gives that $f^{-1} \in Diff^\alpha_b(M)$ for each $f \in Diff^\alpha_b(M)$.

The space $C^\phi_b(U,Y)$ is normed for U bounded in X for $\phi \in \{n,[n]\}$ with $n \in \mathbf{N}$ such that

(1) $\|f\|_{C^n_b(U,Y)} := \sup_{0 \le k \le n; z \in V^{(k)}} \|\bar{\Phi}^k f(z)\|_Y$ or

(2) $\|f\|_{C^{[n]}_b(U,Y)} := \sup_{0 \le k \le n; z \in V^{[k]}} \|\Upsilon^k f(z)\|_Y$,

where $V^{(k)} := \{z \in U^{(k)} : z = (x; v_1, v_2, ...; t_1, t_2, ...), \|v_j\|_X = 1 \forall j\}$, $V^{[k]} := \{z = x^{[k]} \in U^{[k]} : \|v_1^{[k-1]}\|_X = 1, |{}_lv_2^{[q]} t_{q+1}| \le 1, |v_3^{[q]}| \le 1 \quad \forall l,q\}$. The uniformity of $C^\infty_b(U,Y)$ or $C^{[\infty]}_b(U,Y)$ is defined by the family of such norms. Then the uniformity in $C^\alpha_b(M,N)$ is induced by the uniformity in $C^\alpha_b(U,Y)$ by all bounded subsets U in finite unions of ${}_M\phi_i({}_MU_i)$, since to each $f \in C^\alpha_b(M,N)$ there corresponds $f_{j,i} = {}_N\phi_j \circ f \circ {}_M\phi_i^{-1}$ in $C^\alpha_b(U_{j,i},Y)$ with a corresponding domain $U_{j,i} \subset X$. Then application of Formulas A.6(1) or A.7(1) by induction on k and restricting on different embedded subspaces $\mathbf{K}^{[k]}$ or $\mathbf{K}^{(k)}$ in $X^{[k]}$ or $X^{(k)}$ and using uniform continuity on bounded open subsets gives that $(f,g) \mapsto f^{-1} \circ g$ is C^α_b uniformly continuous on bounded subsets of U and $U^{(k)}$ or $U^{[k]}$, where U is a finite union of charts U_j of M.

6. Definition. A topological group G is called a C^α_β Lie group if and only if G has a structure of a $C^{\alpha'}_\beta$ manifold and the mapping $G^2 \ni (f,g) \mapsto f^{-1}g \in G$ is of class of smoothness C^α_β, where $\alpha \le \alpha'$.

7. Theorem. *If M is a C^α_b manifold on $X = c_0(\gamma_X, \mathbf{K})$, where either $\alpha = \infty$ or $\alpha = [\infty]$, then $Diff^\alpha_b(M)$ is the C^α_b Lie group.*

Proof. In view of Theorem 5 it remains to demonstrate that G can be supplied with a structure of C^α_b manifold and that $G^2 \ni (f,g) \mapsto f^{-1}g \in G$ is of class of smoothness C^{α}_b. It is possible to take an equivalent atlas of M consisting of clopen (closed and open simultaneously) charts U_j shrinking it a little in case of necessity. Take a base $\mathcal{W}$ of neighborhoods of id in $Diff^{\alpha}_b(M)$ from the proof of Theorem 5. Then consider a subgroup Ω in $Diff^{\alpha}_b(M)$ such that ΩW covers $Diff^{\alpha}_b(M)$ for each $W \in \mathcal{W}$, where $\Omega W := \{gW : g \in \Omega\}$, $gW := \{gf : f \in W\}$. Therefore, the base $\Omega\mathcal{W}$ generates a topology in $Diff^{\alpha}_b(M)$ equivalent with the initial one (see Chapter 8 in [11]). Moreover, $\Omega\mathcal{W}$ generates a left uniformity in the group of diffeomorphisms.

Consider a subset $W_U := \{f|_U \in W : f \in Diff^{\alpha}_b(M)\}$ for a bounded open U in U_j for some $j \in \Lambda_M$. Then $\phi_i \circ (gW_U) \circ \phi_j^{-1} \subset C^{\alpha}_b(V_j, X)$ and $\phi_i \circ (gW_U) \subset C^{\alpha}_b(U, X)$ for each i, where $V_j := \phi_j(U_j) \subset X$. Therefore, $gW_U \cap hW_V = \{f \in Diff^{\alpha}_b(M) : g^{-1}f|_U \in W, h^{-1}f|_V \in W\}$, $W_U \cap W_V = \{f \in Diff^{\alpha}_b(M) : f|_{U\cup V} \in W\}$. Put $\psi_{i,g,j} := \phi_i \otimes g^{-1} \otimes \phi_j^{-1}$ such that $\psi_{i,g,j}(gW_U) := \phi_i \circ g^{-1}((gW_U) \circ \phi_j^{-1}) \subset C^{\alpha}_b(A, X)$, where $A = \phi_j(U)$, $\psi^{-1}_{l,h,b} := \phi_l^{-1} \otimes h \otimes \phi_b$, hence $\psi^{-1}_{l,h,b} \circ \psi_{i,g,j}(gW_U \cap hW_V) = (\phi_l^{-1} \circ \phi_i) \circ h(W_U \cap g^{-1}hW_V) \circ (\phi_b^{-1} \circ \phi_j)^{-1}$ for $U \subset U_j$, $V \subset U_b$ with $U_j \cap U_b \neq \emptyset$ and $U_l \cap U_i \neq \emptyset$. Thus $\psi_{i,g,j} \circ \psi^{-1}_{l,h,b}$ gives the transition mapping for $Diff^{\alpha}_b(M)$. On the other hand, each $C^{\alpha}_b(A, X)$ has the natural embedding into $C^{\alpha}_b(X, X)$, since X is totally disconnected and A can be taken clopen in X such that each $f \in C^{\alpha}_b(A, X)$ has a $C^{\alpha}_b(X, X)$ extension. For $U_j \cap U_b \neq \emptyset$ the intersection $W_U \cap W_V$ is non void.

Take $At(Diff^{\alpha}_b(M)) := \{(W_{g,U}, \psi_{i,g,j}) : g \in \Omega, i, j \in \Lambda_M\}$ with charts $W_{g,U} := gW_U$, $W \in \mathcal{W}$ and $g \in \Omega$, U bounded in some U_j and with transition mappings $\psi_{i,g,j} \circ \psi^{-1}_{l,h\,b}$ for $W_{h,U}$ and $W_{g,V}$ when $U \cap V \neq \emptyset$. Since $\psi_{i,g,j} \circ \psi^{-1}_{l,h,b} \in C^{\alpha}_b$, then this is the C^{α}_b atlas. The mapping $(f, g) \mapsto f^{-1}g$ is of class $C^{\alpha}_b(Diff^{\alpha}_b(M), Diff^{\alpha}_b(M))$ due to Formulas A.6(1) and A.7(1), since the mappings $(f_{i,j}, g_{i,l}) \mapsto f^{-1}_{i,j} \circ g_{i,l}$ from $C^{\alpha}_b(U, X)$ into $C^{\alpha}_b(V, X)$ with $U \subset \phi_l(U_l) \neq \emptyset$ and $V \subset \phi_j(U_j)$ are of class C^{α}_b.

8. Theorem. *If an ultrametric field* $\mathbf{K}$ *is complete relative to its multiplicative norm and* $\alpha \geq 1$, *then* $Diff^{\alpha}_b(M)$ *is complete as a left uniform space.*

Proof. Recall some facts about uniform spaces. A subset A of the product $S \times S$ of a set S is called a relation in S. The relation inverse to A is denoted $-A$ such that $-A = \{(x, y) : (y, x) \in A\}$ and the composition of relations is denoted $A + B$ such that $A + B = \{(x, z) :$ there exists a $y \in S$ such that $(x, y) \in A$ and $(y, z) \in B\}$. Denote by $\Delta := \{(x, x) : x \in S\}$ the diagonal of the product $S \times S$. Every subset in $S \times S$ containing Δ is called an entourage of the diagonal Δ. The family of all entourages of the diagonal is denoted by $\mathcal{D}_S$. One writes

$|x-y| < V$ if $(x,y) \in V$ and one says that x and y are at a distance less than V. If the condition $|x-y| < V$ is not satisfied, then one writes $|x-y| \geq V$. If $A \subset S$ and $|x-y| < V$ for each $x,y \in A$, then one says that the diameter $\delta(A)$ of A is less than V. Denote $1A := A$, $nA := (n-1)A + A$.

A uniformity $\mathcal{U}$ in a set S is a non-empty subfamily in $\mathcal{D}_S$ satisfying the following four conditions:

$(U1)$ If $V \in \mathcal{U}$ and $V \subset W \in \mathcal{D}_S$, then $W \in \mathcal{U}$;

$(U2)$ If $V_1, V_2 \in \mathcal{U}$, then $V_1 \cap V_2 \in \mathcal{U}$;

$(U3)$ For each $V \in \mathcal{U}$ there exists $W \in \mathcal{U}$ such that $2W \subset V$;

$(U4)$ $\bigcap_{V \in \mathcal{U}} V = \Delta$.

A topological space S is called a T_1 space if for each $x \neq y \in S$ there exists an open set U in S such that $x \in U$ and $y \notin U$. A topological space S is called a Tychonoff space and it is denoted by $T_{3\frac{1}{2}}$ if it is a T_1 space and for each point $x \in S$ and each closed set J in S with $x \notin J$ there exists a continuous function $f : S \to [0,1] \subset \mathbf{R}$ such that $f(x) = 0$ and $f(y) = 1$ for each $y \in J$. If S is a Tychonoff space then for every finite family of functions $f_1, \ldots, f_n \in C^0(S, \mathbf{R})$ or $C^0_b(S, \mathbf{R})$ the formula $\rho_{f_1,\ldots,f_n}(x,y) := \max_{i=1}^n |f_i(x) - f_i(y)|$ defines a pseudo-metric in S. The families of such pseudo-metrics denote by P and P_b respectively. They generate uniformities denoted by $\mathcal{C}$ and $\mathcal{C}_b$ correspondingly (see Chapter 8 in [11]).

If $\mathcal{U}$ is a uniformity in S, then the family $\mathcal{O} := \{U \subset S : \forall x \in U \exists V \in \mathcal{U}$ such that $B(x,V) \subset U\}$ is the topology in S and it is called the topology induced by the uniformity $\mathcal{U}$.

Every covering of S for which there exists $V \in \mathcal{U}$ such that $\mathcal{C}(V)$ is a refinement of it is called a uniform covering relative to $\mathcal{U}$, where $\mathcal{C}(V) := \{B(x,V) : x \in S\}$, $B(x,V) := \{y \in S : |x-y| < V\}$.

If $(S, \mathcal{U})$ is a uniform space and $\mathcal{F}$ is a family of subsets of S, then one says that $\mathcal{F}$ contains arbitrary small sets if for every $V \in \mathcal{U}$ there exists $F \in \mathcal{F}$ such that $\delta(F) < V$. A uniformity $\mathcal{U}$ in S is called complete or $(S, \mathcal{U})$ is called complete, if for each family of closed subsets $\{F_u : u \in \Psi\}$ of the topological space S with the topology induced by $\mathcal{U}$ which has the finite intersection property and contains arbitrary small sets the intersection $\bigcap_{u \in \Psi} F_u$ is non-empty.

If C is an ordered set and $\{x_u : u \in C\}$ is a net in S such that for each $V \in \mathcal{U}$ there exists $u_0 \in C$ for which $|x_u - x_v| < V$ for each $C \ni u, v \geq u_0$, then it is called the Cauchy net. In accordance with theorem of Chapter 8 [11] a uniform space $(S, \mathcal{U})$ is complete if and only if each Cauchy net in this space is convergent. On the other hand, each closed subset A of a complete uniform space $(S, \mathcal{U})$ is complete relative to the uniformity $\mathcal{U}_A$ in A inherited from $\mathcal{U}$ in S.

If G is a topological group and $\mathcal{B}(e) = \mathcal{B}$ is a base of neighborhoods at the unit element e, then each $F \in \mathcal{B}$ determines three coverings of the topological space G: $\mathcal{C}_l(F) = \{xF : x \in G\}$, $\mathcal{C}_r(F) := \{Fx : x \in G\}$, $\mathcal{C}(F) := \{xFy : x, y \in G\}$. By C_l, C_r and C are denoted the families of those coverings of G which have refinements of the form $\mathcal{C}_l(F)$, $\mathcal{C}_r(F)$ or $\mathcal{C}(F)$ respectively. Each of these families generates a uniformity in G. The topology of each of these uniformities is the same as the initial topology in G.

The uniform space $C^{\alpha}_b(M,M)$ is complete for complete M. If U is a bounded canonical closed subset of M contained in a finite number of charts of M, then $Diff^{\alpha}_b(U)$ for each $\alpha \geq 1$ is the neighborhood of $id|_U$ in $C^{\alpha}_b(U,X)$, since from $f \in C^{\alpha}_b(U,X)$ with $\|(id_{i,j} - f_{i,j})|_{\phi_j(U)}\|_{n,U,X} \leq |\pi|$ for each i,j and $1 \leq n \leq \alpha$, $n \in \mathbf{N}$, it follows that $f \in Diff^{\alpha}_b(U)$, where $\pi \in \mathbf{K}$, $0 < |\pi| < 1$, $\| * \|_{n,U,X}$ is either $\| * \|_{C^n_b(U,X)}$ or $\| * \|_{C^{[n]}_b(U,X)}$ for $\alpha \in \mathbf{N} \cup \{\infty\}$ or $\alpha \in \{[1],[2],...\} \cup \{[\infty]\}$ respectively.

Let $\{f_w : w \in C\}$ be a Cauchy net in $Diff^{\alpha}_b(M)$, where C is a directed set. This means that for each neighborhood W of id in $Diff^{\alpha}_b(M)$ there exists $w_0 \in C$ such that $f^{-1}_w f_v \in W$ for each $w_0 \leq w, v \in C$, hence $f_v \in f_{w_0}W$ for each $v \geq w_0$. Since $Diff^{\alpha}_b(M) \subset C^{\alpha}_b(M,M)$ and $C^{\alpha}_b(M,M)$ is complete, then f_v converges in $C^{\alpha}_b(M,M)$ to a function f. Uniformities in the latter space and in the group of diffeomorphisms are related by formulas outlined above. We can take as W a canonical closed subset in $Diff^{\alpha}_b(M)$, $id \in W$. For each bounded U in M as above the restriction $f|_U$ is the diffeomorphism of u onto $f(U)$, $f|_U \in f_{w_0}W_U$.

Relative to the left uniformity in $Diff^{\alpha}_b(M)$ the left shift mapping L_h is uniformly continuous, where $L_h g := hg$ for each h,g in the group. There exists a canonical closed neighborhood U of the unit element id in $C^{\alpha}_b(M,X)$ such that it is contained in the group, since $\alpha \geq 1$ and due to definitions of uniformities in them, while M is modeled on the Banach space. Take a canonical closed symmetric neighborhood V in $Diff^{\alpha}_b(M)$, $V = cl(Int(V))$, $V^{-1} = V$, such that $V^2 \subset U$. If $\{g_n : n\}$ is a Cauchy net in the group, then for each $\varepsilon > 0$ there exists n_0 such that $\rho(g_m, g_n) < \varepsilon$ for all $n,m > n_0$, where ρ is the left-invariant metric in the group. This is equivalent with the fact that for each $\varepsilon > 0$ there exists n_0 so that $\|g^{-1}_n g_m - id\|_{C^{\alpha}_b} < \varepsilon$ in the corresponding space for each $n,m > n_0$. On the other hand, there exists h in the group and $N > 0$ so that $L_h g_m \in V$ for every $m > N$. If X is a Banach space, then from the completeness of $C^{\alpha}_b(M,X)$, and hence of U, since U is closed in $C^{\alpha}_b(M,X)$, it follows, that $Diff^{\alpha}_b(M)$ is also complete as the uniform space (see also Theorems 8.3.6 and 8.3.20 [11]).

Since $\mathbf{K}$ is complete, then X is complete, hence X is the Banach space. Thus $C_b^\alpha(U,X)$ is complete for each U bounded clopen subset in X. Consider a neighborhood W from the base of neighborhoods of id, $W = W_{U,n,\varepsilon,i,j} = \{f \in Diff_b^\alpha(M) : \|f_{i,j}\|_{n,U,X} < \varepsilon\}$, where $i,j \in \Lambda_M$, $n \in \mathbf{N}$, $n \leq \alpha$, U is a clopen bounded subset in X such that $U \subset \phi_j(U_j)$, $0 < \varepsilon < \infty$, $\| * \|_{n,U,X} := \| * \|_{C_b^n(U,X)}$ or $\| * \|_{n,U,X} := \| * \|_{C_b^{[n]}(U,X)}$ respectively. Take without loss of generality $0 < \varepsilon \leq |\pi|$. The manifold M is on the normed space X, hence M is paracompact and it has a locally finite refinement of its atlas (see [11]). Each $\psi_i \circ f_w \circ \phi_j^{-1}$ is converging to a function denoted $f_{i,j}$ and by transfinite induction this consistent family $\{f_{i,j}\}$ induces $f \in C_b^\alpha(M,M)$ such that f_w converges to f. Since $\Upsilon^k \phi_i \circ f_w \circ \phi_j^{-1}(V^{[k]})$ is bounded for each $V^{[k]}$ corresponding to bounded U, then $\Upsilon^k \phi_i \circ f \circ \phi_j^{-1}(V^{[k]})$ is bounded and $\|f_{i,j}\|_{n,U,X} < \infty$, analogously for $\bar{\Phi}^k f$. Thus $f \in C_b^\alpha(M,M)$, consequently, $f \in Diff_b^\alpha(M)$.

May be simply it can be proved using the tangent bundle of the group of diffeomorphisms for $\alpha \geq 1$.

9. Theorem. *If a manifold M has a finite atlas having clopen bounded $\phi_j(U_j) \subset X$, $0 \leq \alpha < \infty$, then $Diff_b^\alpha(M)$ is metrizable by a left invariant metric.*

Proof. The metrization Theorem 8.3 [16] states that if G is a T_0 topological group, then G is metrizable if and only if there is a countable open basis at e. In this case, the metric can be taken left-invariant. If $At(M)$ is finite such that each $\phi_j(U_j)$ is a clopen bounded subset in X, then the base of neighborhoods of id in $Diff_b^\alpha(M)$ is countable and $Diff_b^\alpha(M)$ is metrizable by a left invariant metric in accordance with the general metrization theorem. Practically take as the metric $\rho(f,g) := \rho(id, f^{-1}g) := \max_{i,j \in \Lambda_M} \|id_{i,j} - (f^{-1}g)_{i,j}\|_{C_b^\alpha(V_j,X)}$, where $V_j := \phi_j(U_j)$.

10. Remark. As it is known left and right uniformities in a topological group may be different. Here a left uniformity was considered above. On the same group there may exist several topologies supplying it with structures of a topological group and these topologies need not be comparable.

3. Projective Decomposition of Diffeomorphism Groups

1. Notations and Notes. Let M and N be compact manifolds over a locally compact field $\mathbf{K}$. Suppose that M and N are embedded into $B(\mathbf{K}^{\mathrm{m}},0,1)$ and

$B(\mathbf{K}^{\mathsf{n}},0,1)$ as clopen (closed and open at the same time) subsets [6, 29], where $\mathsf{m},\mathsf{n}\in\mathbf{N}$, $B(X,y,r):=\{z:z\in X;d_X(y,z)\leq r\}$ denotes a clopen ball in a space X with an ultra-metric d_X.

The unit ball $B(\mathbf{K}^{\mathsf{n}},0,1)$ has the ring structure with coordinate wise addition and multiplication, where $char(\mathbf{K})=0$ or $char(\mathbf{K})=p>1$ is a prime number. This ring is isomorphic to a subring of diagonal matrices in the ring $M_n(\mathbf{K})$ of $n\times n$ square matrices over $\mathbf{K}$. Then $B(\mathbf{K}^n,0,|\pi|^k)$ for $k\geq 1$, $\pi\in\mathbf{K}$, $0<|\pi|<1$ is its two-sided ideal, since $\mathbf{K}$ is commutative and $|*|=|*|_{\mathbf{K}}$ is the multiplicative norm in $\mathbf{K}$. Thus there exists the quotient ring $B(\mathbf{K}^n,0,1)/B(\mathbf{K}^n,0,|\pi|^k)$ [4]. The ring $B(\mathbf{K}^{\mathsf{n}},0,1)$ is algebraically isomorphic with the projective limit $B(\mathbf{K}^{\mathsf{n}},0,1)=pr-\lim_k \mathbf{S_{p^k}}^{\mathsf{n}}$, $\mathbf{S_{p^k}}$ is a finite ring consisting of p^{kc} elements such that $\mathbf{S_{p^k}}=\mathbf{S_{p^k}}(\mathbf{K})$ is equal to the quotient ring $B(\mathbf{K},0,1)/B(\mathbf{K},0,p^{-k})$, $\mathbf{S_{p^k}}^{\mathsf{n}}=\mathbf{S_{p^k}}^{\otimes\mathsf{n}}$ is an external product of n copies of $\mathbf{S_{p^k}}$, c is a natural number. Though their structure depends on $char(\mathbf{K})$ we denote these rings by the same symbol depending on $\mathbf{K}$ and omitting it, when a field is specified. In particular $B(\mathbf{F}_{p^n}(\theta),0,1)/B(\mathbf{F}_{p^n}(\theta),0,p^{-k})=(\mathbf{F}_{p^n})^{\otimes p^k}=\mathbf{S_{p^{nk}}}=\mathbf{S_{p^{nk}}}(\mathbf{F}_{p^n}(\theta))$ and $B(\mathbf{Q_p},0,1)/B(\mathbf{Q_p},0,p^{-k})=\mathbf{Z_p}/(p^k\mathbf{Z_p})=\mathbf{S_{p^k}}=\mathbf{S_{p^k}}(\mathbf{Q_p})$ are finite rings consisting of p^{nk} and p^k elements respectively, $\mathbf{Z_p}$ is the ring of p-adic integer numbers, $aB:=\{x:x=ab,b\in B\}$ for a multiplicative group B and its element $a\in B$, $k\in\mathbf{N}$ [38, 43]. The quotient mapping $\pi_k:\mathbf{K}\to\mathbf{K}/B(\mathbf{K},0,p^{-k})$ is defined as well, where $k\in\mathbf{N}$.

Decompositions of continuous and differentiable functions on compact subsets of locally compact fields $\mathbf{K}$ of zero and non-zero characteristics with values in $\mathbf{K}$ into series of polynomials were studied in [1, 40, 7] and references therein. Each function $f\in C^t(M,N)$ has a $C^t(B(\mathbf{K}^{\mathsf{m}},0,1),\mathbf{K}^{\mathsf{n}})$-extension by zero on $B(\mathbf{K}^{\mathsf{m}},0,1)$. Therefore, it has the decomposition

(1) $f=\sum_{l,m} f^l_m \bar{Q}_m e_l$

in the Amice's basis $\bar{Q}_m$, where e_l is the standard orthonormal in the non-archimedean sense [38] basis in $\mathbf{K}^{\mathsf{n}}$ such that $e_l=(0,\dots,0,1,0,\dots)$ with 1 in the l-th place, $m\in\mathbf{Z}^{\mathsf{n}}$, $m_l\geq 0$, $m=(m_1,\dots,m_{\mathsf{n}})$, $f^l_m\in\mathbf{K}$ are expansion coefficients such that $\lim_{l+|m|\to\infty}|f^l_m|_{\mathbf{K}}J(t,m)=0$, $\bar{Q}_m$ are polynomials on $B(\mathbf{K}^{\mathsf{m}},0,1)$ with values in $\mathbf{K}$, $J(t,m):=\|\bar{Q}_m\|_{C^t_b(B(\mathbf{K}^{\mathsf{m}},0,1),\mathbf{K})}$. The space $C^t(M,N)$ is supplied with the uniformity inherited from the space $C^t_b(\mathbf{K}^{\mathsf{m}},\mathbf{K}^{\mathsf{n}})$.

2. Lemma. *Each $f\in C^t(M,N)$ is a projective limit $f=pr-\lim_k f_k$ of polynomials $f_k=\sum_{l,m} f^l_{m,k}\bar{Q}_{m,k}e_l$ on rings $\mathbf{S}^{\mathsf{m}}_{\mathbf{p^k}}=\mathbf{S}^{\mathsf{m}}_{\mathbf{p^k}}(\mathbf{K})$ with values in $\mathbf{S}^{\mathsf{n}}_{\mathbf{p^k}}=\mathbf{S}^{\mathsf{n}}_{\mathbf{p^k}}(\mathbf{K})$, where $f^l_{m,k}\in\mathbf{S_{p^k}}$ and $\bar{Q}_{m,k}$ are polynomials on $\mathbf{S}^{\mathsf{m}}_{\mathbf{p^k}}$ with values in $\mathbf{S_{p^k}}$.*

Proof. For each $m \geq k$ consider the quotient mappings (ring homomorphisms): $\pi_m : B(\mathbf{K}, 0, 1) \to \mathbf{S}_{\mathbf{p}^m}$ and $\pi_k^m : \mathbf{S}_{\mathbf{p}^m} \to \mathbf{S}_{\mathbf{p}^k}$ (see §1). This induces the quotient mappings $\pi_m : N \to N_m$ and $\pi_k^m : N_m \to N_k$, where $N_m \subset \mathbf{S}_{\mathbf{p}^m}$, $\pi_k^m \circ \pi_m = \pi_k$, $\pi_k^k = id_k : \mathbf{S}_{\mathbf{p}^k} \to \mathbf{S}_{\mathbf{p}^k}$.

Let now M and N be two analytic compact manifolds embedded into $B(\mathbf{K}^{\mathrm{m}}, 0, 1)$ and $B(\mathbf{K}^{\mathrm{n}}, 0, 1)$ respectively as clopen subsets and $f \in C^t(M,N)$, where $C^t(M,N)$ denotes the space of functions $f : M \to N$ of class C^t, $t \geq 0$. There exists $s \in \mathbf{N}$ such that if $x \in M$ and $y \in N$, then $B(\mathbf{K}^{\mathrm{m}}, x, p^{-s}) \subset M$ and $B(\mathbf{K}^{\mathrm{n}}, y, p^{-s}) \subset N$. Therefore, consider the cofinal set $\Lambda_s := \{k : k \geq s, k \in \mathbf{N}\}$ in $\mathbf{N}$. For an integer t it is a space of t-times continuously differetiable functions in the sense of partial difference quotients (see Section 2 and [21, 23, 40]). Thus $f = pr - \lim_k f_k$, where $f_k := \pi_k^*(f)$, π_k^* is naturally induced by π_k using the polynomial expansion of f (see in details below), where such decomposition exists for each continuous $f : M \to N$ due to Formula 1(1) (for the limit of an inverse sequence, see [4], §2.5 [11] and §§3.3, 12.202 [32]). Put $C^t(M_k, N_k) := \pi_k^* \circ C^t(M,N) = \{f_k : f \in C^t(M,N)\}$, hence

(1) $C^t(M,N) \subset pr - \lim_k C^t(M_k, N_k)$

algebraically without taking into account topologies. Thus write it in the form:

(2) $C^t(M,N) = T - pr - \lim\{C^t(M_k,N_k), \pi_l^k, \Lambda_s | C^t(M,N)\}$ algebraically, where

(3) $T - pr - \lim\{P_k, \pi_l^k, \Lambda | G\} := \{f : pr - \lim\{f_k, \pi_l^k, \Lambda\} = f, f \in G, f_k \in P_k \ \forall k \in \Lambda\}$

denotes the conditional projective limit with a condition G, since $T - pr - \lim\{P_k, \pi_l^k, \Lambda | G\} = G \cap pr - \lim\{P_k, \pi_l^k, \Lambda\}$.

Indeed, in accordance with §1 $f_k = \pi_k^*(f)$ and

(4) $\pi_k^*(f(x)) = \sum_{l,m} (\pi_k(f_m^l)(\pi_k^* \bar{Q}_m(x)) e_l$,

since π_k is the ring homomorphism and $\pi_k(e_l) = e_l$. Then $\pi_k(ax^m) = a_k x^m(k)$ for each $a \in \mathbf{K}$ and $x \in B(\mathbf{K}^{\mathrm{m}}, 0, 1)$, where $x^m = x_1^{m_1} \dots x_{\mathrm{m}}^{m_{\mathrm{m}}}$, $a_k = \pi_k(a)$ with $a_k \in \mathbf{S}_{\mathbf{p}^k}$ and $x^m(k) := \pi_k(x^m)$ with $x(k) \in \mathbf{S}_{\mathbf{p}^k}^{\mathrm{m}}$, hence

(5) $\pi_k^*(\bar{Q}_m(x)) = \bar{Q}_{m,k}(x(k))$.

If $\bar{Q}_m(x) = \sum_{q, 0 \leq q_j \leq m_j \forall j} b_q x^q$, where $q = (q_1, \dots, q_{\mathrm{m}})$, $m = (m_1, \dots, m_{\mathrm{m}})$, $m_j \in \{0, 1, 2, \dots\} =: \mathbf{N_0}$, $j = 1, \dots, \mathrm{m}$, $b_q \in \mathbf{K}$, $x^q := x_1^{q_1} \dots x_{\mathrm{m}}^{q_{\mathrm{m}}}$, then

$\bar{Q}_{m,k}(x(k)) = \sum_{q, 0 \leq q_j \leq m_j \forall j} \pi_k(b_q) x(k)^q$,

$x(k) = (\pi_k(x_1), \dots, \pi_k(x_{\mathrm{m}}))$, since $\pi_k : \mathbf{K} \to \mathbf{K}/B(\mathbf{K}, 0, p^{-k})$ is defined as well. The series for f_k is finite, since $\pi_k(a) = 0$ for each $a \in \mathbf{K}$ with $|a| < p^{-k}$ and

(6) $\lim_{l+|m| \to \infty} |f_m^l|_{\mathbf{K}} J(t,m) = 0$. Therefore,

(7) $f(x) = pr - \lim\{f_k(x(k)), \pi_l^k, \Lambda_s\}$ for each $x \in M$.

As shows this proof the index t can be omitted from $C^t(M_k, N_k)$, where $k \in \Lambda_s$. More precisely we have the following corollary.

3. Corolary. *The space $C^t(M_k, N_k)$ is independent from t and it is algebraically isomorphic with the space $N_k^{M_k}$ of all mappings from M_k into N_k for each $k \in \Lambda_s$. Moreover, $(\mathbf{S}^{\mathrm{n}}_{\mathbf{p}^{\mathrm{k}}})^{(\mathbf{S}^{\mathrm{m}}_{\mathbf{p}^{\mathrm{k}}})}$ is a finite-dimensional space over the ring $\mathbf{S}_{\mathbf{p}^{\mathrm{k}}}$.*

Proof. In view of Lemma 2 in the module $C^t(\mathbf{S}^{\mathrm{m}}_{\mathbf{p}^{\mathrm{k}}}, \mathbf{S}_{\mathbf{p}^{\mathrm{k}}})$ of the ring $\mathbf{S}_{\mathbf{p}^{\mathrm{k}}}$ there is only a finite number of $\mathbf{S}_{\mathbf{p}^{\mathrm{k}}}$-linearly independent polynomilas $\bar{Q}_{m,k}(x(k))$, since the rings $\mathbf{S}^{\mathrm{m}}_{\mathbf{p}^{\mathrm{k}}}$ and $\mathbf{S}_{\mathbf{p}^{\mathrm{k}}}$ are finite, also $z^a = z^b$ for each natural numbers a and b such that $a = b \ (mod\ (p^k))$ and each $z \in \mathbf{S}_{\mathbf{p}^{\mathrm{k}}}$. The space $C^t(M_k, N_k)$ is isomorphic with $N_k^{M_k}$, since M_k and N_k are discrete.

Therefore, denote $C^t(M_k, N_k)$ by $C(M_k, N_k)$.

4. Corollary. *There exists the group $\pi_k^* \circ Diff^{\alpha}(M)$ isomorphic with the symmetric group $\Sigma_{\mathrm{n_k}}$ for every $k \in \Lambda_s$, where $\mathrm{n_k}$ is the cardinality of M_k, $\alpha \in \{n, [n] : 0 \le n \le \infty\}$.*

Proof. For each $k \in \Lambda_s$ there exists the mapping $\pi_k : M \to M_k$ (see the proof of lemma 2). Since π_k is the quotient continuous mapping, then $\pi_k(M) = M_k$. If $f : M \to M$ is a continuous epimorphism, $f(M) = M$, then $\pi_k(f(M)) = M_k$. Let $z \in M_k$, $k \in \Lambda_s$, then there exists $x \in M$ such that $\pi_k(x) = z$, hence $f_k(z) = \pi_k^*(f(\pi_k^{-1}(z)))$ due to Formulas 2(4,5), since $\pi_k(\pi_k^{-1}(z)) = z$ and $\pi_k(x) = x(k) = z$. Then $\pi_k^{-1}(M_k) = M$ and $\pi_k^{*-1}(f_k(M_k)) = \bigcup_{z \in M_k}\{f(x) : x \in \pi_k^{-1}(z)\}$, consequently, $\pi_k^{*-1}(f_k(M_k)) = M$, since for each $x \in M$ there exists $z \in M_k$ such that $\pi_k(x) = z$. Therefore, if $h \in Diff^{\alpha}(M)$, then $h_k(M_k) = M_k$, since $h(M) = M$ and M_k is finite. In accordance with Theorem 3.2.14 [11] if $\{\phi, f_{\sigma'}\}$ is a mapping of an inverse spectrum $\{X_\sigma, \pi_\rho^\sigma, \Psi\}$ of compacts into an inverse spectrum $\{Y_{\sigma'}, \pi_{\rho'}^{\sigma'}, \Psi'\}$ of T_1-spaces and all $f_{\sigma'}$ are epimorphisms on $Y_{\sigma'}$, then the limit mapping $f = pr - \lim\{\phi, f_{\sigma'}\}$ is also the epimorphism. In view of Corollary 3 $\pi_k^* \circ Diff^{\alpha}(M)$ is algebraically isomorphic with the following discrete group $Hom(M_k)$ of all homeomorphisms h_k of M_k, that is, bijective surjective mappings $h_k : M_k \to M_k$. Using an enumeration of elements of M_k we get an isomorphism of $Hom(M_k)$ with $\Sigma_{\mathrm{n_k}}$.

5. Suppose that $C_w(M, N) := pr - \lim_k\{N_k^{M_k}, \pi_l^k, \Lambda_s\}$ is an uniform space of continuous mappings $f : M \to N$ supplied with an uniformity and projective weak topology as well inherited from products of uniform spaces $\prod_{k=1}^{\infty} N_k^{M_k}$ (see also §8.2 [11] and §1 above). Denote the corresponding projective weak topology in $C_w(M,N)$ by τ_w. The spaces $C^\alpha(M,N)$ and $C_w(M,N)$ are subsets of **K**-linear spaces $C^\alpha(M,\mathbf{K}^n)$ and $C^0(M,\mathbf{K}^n)$ respectively. Supply with algebraic structures subsets of the latter **K**-linear spaces as inherited from them.

Corollary. *The uniform space $C^\alpha(M,N)$ is not algebraically isomorphic with $C_w(M,N)$, when $\alpha > 0$. The topological space $C_w(M,N)$ is compact.*

Proof. In accordance with appendix in [4] and §2.5 [11] and Formulas $2(1-6)$ above the spaces $C^0(M,N)$ and $C_w(M,N)$ coincide algebraically, since the connecting mappings π_n^m are uniformly continuous for each $m \geq n$. The space $C^0(M,\mathbf{K}^n)$ is **K**-linear and its uniformity is completely determined by a neighborhood base of zero. The space $C_w(M,N)$ is uniformly homeomorphic with $pr-\lim_k(\mathbf{S}_{\mathbf{p}^k})^{M_k}$, which is compact by the Tychonoff Theorem 3.2.4 [11]. Since $C^0(M,N) \neq C^\alpha(M,N)$ for $\alpha > 0$, then $C_w(M,N)$ and $C^\alpha(M,N)$ are different algebraically see Formulas $2(1-3)$).

Remark. In general in two consequtive projective limits of topological spaces two limits may be non commuting.

6. Suppose that $Diff_w(M) := pr-\lim_k Hom(M_k)$ is supplied with the uniformity inherited from $C_w(M,M)$. The group $Diff_w(M)$ is called the non-archimedean compactification of $Diff^\alpha(M)$.

Theorem. *The group $Diff_w(M)$ is the compact topological group and it is the compactification of $Diff^\alpha(M)$ in the projective weak topology τ_w. If $\alpha > 0$, then $Diff^\alpha(M)$ does not coincide with $Diff_w(M)$.*

Proof. From $Diff^\alpha(M) \subset C^\alpha(M,M)$ it follows that $Diff^\alpha(M)$ has the corresponding algebraic embedding into $C_w(M,M)$ as the set. Since $C_w(M,M)$ is compact and $Hom(M)$ is a closed subset in $C_w(M,M)$, then due to Corollary 5 $Hom(M) \cap C_w(M,M) = Diff_w(M)$ is compact. The topological space $C^\alpha(M,M)$ is dense in $C^0(M,M)$, consequently, $Diff^\alpha(M)$ is dense in $Diff_w(M)$. If $\alpha > 0$, then $Diff^\alpha(M) \neq Hom(M)$, hence $Diff^\alpha(M)$ and $Diff_w(M)$ do not coincide algebraically. It remains to verify, that $Diff_w(M)$ is the topological group in its projective weak topology τ_w. If $f,g \in C^\alpha(M,N)$, then $\pi_k^*(\bar{Q}_m(g(x))) = Q_{m,k}(g_k(x(k))$ due to Formula 2(4), consequently, $\pi_k^*(f \circ g) = \sum_{l,m} \pi_k(f_m^l) Q_{m,k}(g_k(x(k))e_l$, hence

(1) $(f \circ g)_k = f_k \circ g_k$.

Since $\pi_k(x) = x(k)$, then $\pi_k^*(id(x)) = id_k(x(k))$, where $id(x) = x$ for each $x \in M$. Therefore, for $f = g^{-1}$ we have $(f \circ g)_k = f_k \circ g_k = id_k$, hence

(2) $\pi_k^*(g^{-1}) = g_k^{-1}$.

The associativity of the composition $(f_k \circ g_k) \circ h_k = f_k \circ (g_k \circ h_k)$ of all functions $f_k, g_k, h_k \in Hom(M_k)$ together with others properties given above means, that $Diff_w(M)$ is the algebraic group.

Indeed, inverse limits of mappings $f = pr - \lim_k f_k$, $g = pr - \lim_k g_k$ and $h = pr - \lim_k h_k$ satisfy the associativity axiom as well, each f has the inverse element $f^{-1} = pr - \lim_k f_k^{-1}$ such that $f^{-1}(f(x)) = id$ and $e = id$ is the unit element. By the definition of the weak topology in $Diff_w(M)$ for each neighborhood of $e = id$ in $Diff_w(M)$ there exists $k \in \mathbf{N}$ and a subset $W_k \subset Hom(M_k)$ such that $e_k \in \pi_k^{-1}(W_k) \subset W$. On the other hand, $Hom(M_k)$ is discrete, hence there are $e_k \in V_k \subset Hom(M_k)$ and $e_k \in U_k \subset Hom(M_k)$ such that $V_k U_k \subset W_k$, hence there are neighborhoods $e \in V \subset Diff_w(M)$ and $e \in U \subset Diff_w(M)$ such that $VU \subset W$, where $V = \pi_k^{-1}(V_k)$, $U = \pi_k^{-1}(U_k)$ and $VU = \{h : h = f \circ g, f \in V, g \in U\}$. Consider a neighborhood W' of f^{-1}, then $V := W' f^{-1}$ is the neighborhood of e and there exists $k \in \mathbf{N}$ such that $\pi_k^{-1}(e_k) =: U \subset V^{-1}$, since $e_k^{-1} = e_k$ and π_k is the homomorphism. Thus, $fU := W$ is the neighborhood of f such that $W^{-1} \subset W'$, hence the inversion operation $f \mapsto f^{-1}$ is continuous.

7. Theorem. *The initial C^α topology τ_α and the weak projective limit topology τ_w in $Diff^\alpha(M)$ are incomparable, where M is compact.*

Proof. Remind that a topology $\mathcal{O}_1$ in a topological space F is called weaker, than a topology $\mathcal{O}_2$, if $\mathcal{O}_1 \subset \mathcal{O}_2$ or one says that $\mathcal{O}_2$ is stronger than $\mathcal{O}_1$. Up to a diffeomorphism as above consider M clopen in $B(\mathbf{K}^n, 0, 1)$. Since $Diff^\beta(M)$ is contained in $Diff^\alpha(M)$ for each $\beta > \alpha \geq 0$ and τ_β is stronger than τ_α in $Diff^\beta(M)$, then it is sufficient to prove this theorem for $\alpha \geq 1$. Each projection $\pi_k^* : C^\alpha(M, \mathbf{K}^n) \to (\mathbf{K}_\mathbf{k}^n)^{M_k}$ induces the quotient metric ρ_k in the $\mathbf{K}_\mathbf{k}$-module $(\mathbf{K}_\mathbf{k}^n)^{M_k}$ such that $\rho_k(f_k, g_k) := \inf_{z, \pi_k(z)=0} \|f - g + z\|_{C^\alpha(M, \mathbf{K}^n)}$, where $\mathbf{K}_\mathbf{k} := \mathbf{K}/B(\mathbf{K}, 0, p^{-k})$ is the quotient ring and π_k is induced by such quotient mapping from $\mathbf{K}$ onto $\mathbf{K}_\mathbf{k}$.

In view of the Kaplansky theorem which is the non-archimedean analog of the Stone-Weierstrass theorem and true over fields of zero and non-zero characteristics in
$C^\alpha(B(\mathbf{K}^n, 0, 1), \mathbf{K})$ the set of polynomials is dense [40, 7]. Thus after application of the quotient mapping we get that in the module $C(\mathbf{S}_{\mathbf{p}^\mathbf{k}}{}^n, \mathbf{S}_{\mathbf{p}^\mathbf{k}})$ over the finite ring $\mathbf{S}_{\mathbf{p}^\mathbf{k}}$ the set of polynomials is dense. Moreover, each continuous function has a decomposition into converging series of polynomials. The locally compact field $\mathbf{K}$ is commutative [43], hence the ring $\mathbf{S}_{\mathbf{p}^\mathbf{k}}$ is commutative. This

means that the multiplicative group ${\mathbf{S}_{\mathbf{p}^k}}^* := \mathbf{S}_{\mathbf{p}^k} \setminus \{0\}$ is commutative and consists of $p^k - 1$ elements. Thus, if $0 \neq x \in \mathbf{S}_{\mathbf{p}^k}$, then $x^{p^k} = 1$. Therefore, over $\mathbf{S}_{\mathbf{p}^k}$ the set of polynomials is finite dimensional, hence each $f_k \in C({\mathbf{S}_{\mathbf{p}^k}}^{\mathsf{n}}, \mathbf{S}_{\mathbf{p}^k})$ is polynomial over $\mathbf{S}_{\mathbf{p}^k}$.

Again consider $C^\alpha(B(\mathbf{K}^{\mathsf{n}},0,1),\mathbf{K})$, which is the algebra over $\mathbf{K}$. If $f_i(x) = x_i$, where $x = (x_1,\dots,x_{\mathsf{n}}) \in B(\mathbf{K}^{\mathsf{n}},0,1)$, then $\prod_{i=1}^{\mathsf{n}} f_i(x)^{s_i} = \prod_{i=1}^{\mathsf{n}} x_i^{s_i} \in C^\alpha(B(\mathbf{K}^{\mathsf{n}},0,1),\mathbf{K})$, where $x_i^s := \prod_{j=1}^s (x_i)_j$ with $(x_i)_j = x_i$ for each j, where $s, s_i \in \mathbf{N}$. In particular, it contains a subalgebra $\mathcal{A}_k$ containing all constants from $\mathbf{K}$ and all polynomials of the form

(1) $\sum_{k_1=p^k l(1),\dots,k_{\mathsf{n}}=p^k l(\mathsf{n})} a_{k_1,\dots,k_{\mathsf{n}}} x_1^{k_1} \dots x_{\mathsf{n}}^{k_{\mathsf{n}}}$,

where $a_{k_1,\dots,k_{\mathsf{n}}} \in \mathbf{K}$, $l(i) \in \{0,1,2,\dots\}$ with

(2) $\sum_{k_1=p^k l(1),\dots,k_{\mathsf{n}}=p^k l(\mathsf{n})} a_{k_1,\dots,k_{\mathsf{n}}} = 0$, $1 \leq k \in \mathbf{Z}$.

This algebra $\mathcal{A}_k$ over $\mathbf{K}$ separates points in $C^\alpha(B(\mathbf{K}^{\mathsf{n}},0,1),\mathbf{K})$ and by the Kaplansky theorem $\mathcal{A}_k$ is everywhere dense in $C^\alpha(B(\mathbf{K}^{\mathsf{n}},0,1),\mathbf{K}))$. But $\pi_k(x_i^{k_i}) = (\pi_k(x_i^{p^k}))^{l(i)} = 1$ for each i and $k_i = p^k l(i)$ with $l(i) \in \{0,1,2,\dots\}$. Consider a polynomial with values in $\mathbf{K}^{\mathsf{n}}$ of the form

(3) $f = id + \sum_{i=1}^{\mathsf{n}} \sum_{k_1=p^k l(1),\dots,k_{\mathsf{n}}=p^k l(\mathsf{n})} a_{i,k_1,\dots,k_{\mathsf{n}}} x_1^{k_1} \dots x_{\mathsf{n}}^{k_{\mathsf{n}}} e_i$,

where $(f - id) \in \mathcal{A}_k^{\mathsf{n}}$, $|a_{i,k_1,\dots,k_{\mathsf{n}}}| \leq |\pi|$ for each $i, l(1),\dots,l(\mathsf{n})$, where $\pi \in B(\mathbf{K},0,1)$, $|\pi| < 1$, $|\pi| = \max\{y \in \Gamma_{\mathbf{K}} : 0 < y < 1\}$. Then $\|f - id\|_{C^\alpha(B(\mathbf{K}^{\mathsf{n}},0,1),\mathbf{K})} \leq |\pi|$, consequently, f is the isometry and $f \in Diff^\alpha(M)$ and inevitably

$f_k := \pi_k^*(f) = \pi_k^*(id) + \sum_{i=1}^{\mathsf{n}} \sum_{k_1=p^k l(1),\dots,k_{\mathsf{n}}=p^k l(\mathsf{n})} \pi_k(a_{i,k_1,\dots,k_{\mathsf{n}}}) e_i = \pi_k^*(id)$

due to Condition (2). Therefore, $(\pi_k^*)^{-1}(e_k)$ is everywhere dense in a neighborhood of $e = id$ in $Diff^\alpha(M)$, where $e_k \in \pi_k^*(Diff^\alpha(M)) = Hom(M_k)$ is the unit element, $k > 1$, $card(M_k) > 1$, $Hom(M_k)$ is the symmetric group of M_k elements.

On the other hand, there is $\mathsf{c} = card(\mathbf{R})$ elements $f \in Diff^\alpha(M)$ with $\|f - id\|_{C^\alpha(B(\mathbf{K}^{\mathsf{n}},0,1),\mathbf{K})} \leq |\pi|$ such that
$f = id + \sum_{i=1}^{\mathsf{n}} \sum_{k_1=p^k l(1),\dots,k_{\mathsf{n}}=p^k l(\mathsf{n})} a_{i,k_1,\dots,k_{\mathsf{n}}} x_1^{k_1} \dots x_{\mathsf{n}}^{k_{\mathsf{n}}} e_i$, but with
$\sum_{k_1=p^k l(1),\dots,k_{\mathsf{n}}=p^k l(\mathsf{n})} a_{i,k_1,\dots,k_{\mathsf{n}}} \neq 0$ for which $\pi_k^*(f) \neq e_k$.

Thus the set $\pi_k^{*-1}(e_k)$ is open in $(Diff^\alpha(M), \tau_w)$, but $\pi_k^{*-1}(e_k) \notin \tau_\alpha$. At the same time, $\{f \in Diff^\alpha(M) : \|f - id\|_{C^\alpha(B(\mathbf{K}^{\mathsf{n}},0,1),\mathbf{K}^{\mathsf{n}})} \leq |\pi|\}$ is open in $(Diff^\alpha(M), \tau_\alpha)$, but it is not open in τ_w topology. This proves the assertion of this theorem, since neither τ_α nor τ_w is weaker among two of them and they are different.

8. Theorem. *Let M be a C^α manifold finite dimensional over a non-archimedean infinite field $\mathbf{K}$ with a non trivial multiplicative norm complete relative to its uniformity, where $\mathbf{K}$ may be non locally compact. Suppose that $Diff_b^\alpha(M)$ is the group of all uniformly C^α continuous diffeomorphisms of M, where M is embedded into $\mathbf{K}^n$ as the bounded clopen subset, $\alpha \in \{t, [t]\}$, $1 \le t \le \infty$. Then $G_s := \{g \in Diff_b^\alpha(M) : \|g - id\|_{C^\alpha(M,\mathbf{K}^n)} < |\pi|^s\}$ has a non-archimedean completion, which is a group, where $\pi \in \mathbf{K}$, $|\pi| < 1$, $1 \le s \in \mathbf{N}$.*

Proof. If $g \in G_s$, then g is the isometry: $|g(x) - g(y)| = |x - y|$ for each $x, y \in M \subset \mathbf{K}^n$, since $1 \le t$ and $g(x) - g(y) = \Phi^1 g(x; x - y; 1)$, where $1 \le s$. Without loss of generality up to an affine diffeomorphism of $\mathbf{K}^n$ we can consider, that $M \subset B(\mathbf{K}^n, 0, 1)$. Consider the ring homomorphism $\pi_k : B(\mathbf{K}^n, 0, 1) \to B(\mathbf{K}^n, 0, 1)/B(\mathbf{K}^n, 0, |\pi|^k)$, where the quotient ring is discrete and can be supplied with the quotient norm. Then $\pi_k(M) =: M_k$ is the discrete topological space.

Since M is clopen in $\mathbf{K}^n$ then each $g \in Diff_b^\alpha(M)$ has the extension as *id* on $\mathbf{K}^n \setminus M$. Therefore, without loss of generality consider M such that there exists $q \in \mathbf{N}$ for which from $x \in M$ it follows that $B(\mathbf{K}^n, x, |\pi|^q) \subset M$. Consider the cofinal set $\Lambda_q := \{k \in \mathbf{N} : k \ge q\}$. If $g \in G_s$, then

(1) $g(B(\mathbf{K}^n, x, |\pi|^k)) = B(\mathbf{K}^n, g(x), |\pi|^k)$ for each $B(\mathbf{K}^n, x, |\pi|^k) \subset M$, since g is the isometry. Consequently,

(2) $f \circ g(B(\mathbf{K}^n, x, |\pi|^k)) = f(B(\mathbf{K}^n, g(x), |\pi|^k)) = B(\mathbf{K}^n, f(g(x)), |\pi|^k)$ and $g^{-1}(B(\mathbf{K}^n, x, |\pi|^k)) = B(\mathbf{K}^n, g^{-1}(x), |\pi|^k)$ for each $f, g \in G_s$.

Therefore, g and π_k generate the natural mapping ${}_kg : M_k \to M_k$ such that it is bijective and epimorphic, since $\pi_k^{-1}(z) = B(\mathbf{K}^n, x, |\pi|^k)$ for each $z \in \pi_k(B(\mathbf{K}^n, 0, 1))$, and

(3) $B(\mathbf{K}^n, x, |\pi|^k) = x + B(\mathbf{K}^n, 0, |\pi|^k)$, where $x \in B(\mathbf{K}^n, 0, 1)$ is such that $\pi_k(x) = z$, $\pi_k^*(g(x)) = {}_kg(z) := \pi_k \circ g \circ \pi_k^{-1}(z)$ for each $z \in M_k$. Thus due to Equations $(1-3)$ there exists the discrete group $\pi_k^*(G_s) =: {}_kG_s$ and $\pi_k^l({}_lG_s) = {}_kG_s$ for each $l \ge k$, where π_k^l are mappings of the inverse system such that $\pi_k^l \circ \pi_l = \pi_k$ such that π_k^l and π_l, π_k are algebraic homomorphisms, π_k^l are written without star for simplicity of notation. As in Theorem 6 it gives the inverse sequence of discrete groups $S = \{{}_lG_s, \pi_k^l, \Lambda_q\}$. In view of Lemma 2.5.9 [11] if $\{\phi, f_{\sigma'}\}$ is a mapping of an inverse system $S = \{X_\sigma, \pi_\rho^\sigma, \Psi\}$ into an inverse system $S' = \{Y_{\sigma'}, \pi_{\rho'}^{\sigma'}, \Psi'\}$ and all mappings $f_{\sigma'}$ are injective, then the limit mapping $f = \lim\{\phi, f_{\sigma'}\}$ is also injective. If moreover, all $f_{\sigma'}$ are surjective, then f is also surjective.

Each discrete topological group is complete relative to its left uniformity generated by its topology. Thus, the limit $\lim S$ of the inverse system of discrete groups is the Tychonoff topological group relative to the projective weak topology inherited from the product Tychonoff topology τ_w, $G \subset \lim S \subset \prod_{l\in\Lambda_q} {}_lG_s$. Moreover, $\lim S =: G_s^w$ is the complete uniform space with the left uniformity $\mathcal{T}_w$ generated by the left shifts and the neighborhood base of e in G_s^w, since each ${}_kG_s$ is complete (see Theorems 2.5.13, 8.3.6 and 8.3.9 [11]). We have that algebraically $G_s \subset G_s^w$, G_s is the topological group relative to the topology inherited from G_s^w. In view of Theorem 7 the $\tau_w|_{G_s}$ topology is incomparable with the C_b^α uniformly bounded continuous topology.

9. Corollary. *Let $Diff_b^\alpha(M)$ be the group as in Theorem 8. Then $Diff_b^\alpha(M)$ has the non-archimedean completion which is the topological group.*

Proof. Let G_s be the subgroup as in Theorem 8, then G_s is clopen in $Diff_b^\alpha(M)$. There exists a family $g_j \in Diff_b^\alpha(M)$ such that $\bigcup_{j\in\Psi} g_jG_s = Diff_b^\alpha(M)$, where Ψ is a set. We have that gG_s is clopen in $Diff_b^\alpha(M)$ for each $g \in Diff_b^\alpha(M)$, since $L_g : Diff_b^\alpha(M) \to Diff_b^\alpha(M)$ is the homeomorphism, where $L_g(h) = gh$ for each $g,h \in Diff_b^\alpha(M)$. Then $g_jG_s \cap g_iG_s = g_j(G_s \cap g_j^{-1}g_iG_s)$. Let $g_{j_0} = id$ and $g_j \neq id$ for each $j \neq j_0$.

The group $Diff_b^\alpha(M)$ is metrizable, since $C_b^\alpha(M,\mathbf{K}^n) \subset C_b^\alpha(B(\mathbf{K}^n,0,1),\mathbf{K}^n)$ is metrizable, hence it is paracompact. Each two balls in $C_b^\alpha(M,\mathbf{K}^n)$ either are disjoint or coincide, since it is the normed space. We have that $Diff_b^\alpha(M)$ is contained in $C_b^\alpha(M,\mathbf{K}^n)$ and at the same time $Diff_b^\alpha(M)$ is the neighborhood of id in $C_b^\alpha(M,\mathbf{K}^n)$, $B(C_b^\alpha(M,\mathbf{K}^n),id,|\pi|^s) = G_s$ for $s \geq 1$ and $1 \leq t$, $\alpha \in \{t,[t]\}$. Therefore choose Ψ such that $g_jG_s \cap g_iG_s = \emptyset$ for each $i \neq j \in \Psi$, hence $g_i^{-1}g_j \notin G_s$ for each $i \neq j$.

The minimal group $gr(\bigcup_{j\in\Psi} g_jG_s) = Diff_b^\alpha(M)$ and it is contained in the minimal algebraic group $gr(\bigcup_{j\in\Psi} g_jG_s^w) =: G^w$. Supply the latter group with the uniformity $\mathcal{T}_w$ induced from the base of neighborhoods of e in G_s^w with the help of left shifts. Then the restriction of $\mathcal{T}_w$ on G_s^w coincides with that of $\mathcal{T}_w$ in Theorem 8. Since G_s^w is clopen in G^w and G_s^w is complete, then for each Cauchy net $\{h_q : q \in \nu\}$ in G^w, where ν is an ordered set, there exists q_0 such that for each $q,l \geq q_0$ there is the inclusion $h_q^{-1}h_l \in G_s^w$, hence $\{h_{q_0}^{-1}h_l : l > q_0\}$ converges in G_s^w to some $g_s \in G_s^w$, since G_s^w is complete, consequently, $\{h_q : q \in \nu\}$ converges in G^w to $h_{q_0}g_s \in G^w$ and inevitably G^w is complete (see also Theorem 8.3.20 [11]). This G^w is the desired non-archimedean completion.

4. Example of the Group of Diffeomorphisms

This section contains the example of the group of diffeomorphisms. It illustrates the general theory. For the group of diffeomorphisms of $\mathbf{Z_p}$ of class C^t, where $0 \le t \le \infty$, formulas for expansion coefficients in the Mahler base of compositions $g \circ f$ and inverse elements f^{-1} of diffeomorphisms f and g are found.

Let C^t be a class of smoothness of functions $f : M \to \mathbf{K}$ as in §1 (see also [1, 40, 21, 22]), where M is a Banach manifold over a complete (as an uniform space) non-Archimedean infinite field $\mathbf{K}$ with non-trivial valuation and of zero characteristic $char(\mathbf{K}) = 0$. Suppose that $L_b f(x) := f(x+b)$ is a shift operator, $\Delta_b f(x) := (L_b - I)f(x)$ is a difference operator, $L^0 = I, \Delta^0 = I, L := L_1, \Delta := \Delta_1$, where $x, b \in \mathbf{K}$. For a product of two functions $f, g : \mathbf{K} \to \mathbf{K}$ there are formulas:

(1) $\Delta[f(x)g(x)] = (\Delta f)(x)(Lg)(x) + f(x)(\Delta g)(x)$ and

(2) $L\Delta = \Delta L$. Therefore,

$\Delta^k[f(x)g(x)] = \sum_{j=0}^{k} \binom{k}{j} [\Delta^j f(x)] L^j [\Delta^{k-j} g(x)]$ for each $k \in \mathbf{N}$, consequently,

(3) $\Delta^k f_1(x) \dots f_n(x) = \sum_{k_1+\dots+k_n=k} (k!/(k_1!\dots k_n!)) [\Delta^{k_1} f_1(x)] L^{k_1} [\Delta^{k_2} f_2(x)] L^{k_2} \dots [\Delta^{k_{n-1}} f_{n-1}(x)] L^{k_{n-1}} [\Delta^{k_n} f_n(x)]$, where L^{k_j} acts on all functions situated on the right from it. If $f \in C^t(\mathbf{Z_p}, \mathbf{Q_p})$, then there is its expansion in the Mahler base $\binom{x}{n}$ as a series:

(4) $f(x) = \sum_{j=0}^{\infty} f_j \binom{x}{j}$, where $f_j \in \mathbf{Q_p}$,

(5) $\lim_{j\to\infty} |f_j| j^t = 0$ for $0 \le t < \infty$ and

(6) $f_j = [\Delta^j f(x)]|_{x=0}$, since

(7) $\Delta \binom{x}{j} = \binom{x}{j-1}$ for each $j \in \mathbf{N}$, $\binom{x}{0} = 1$ and $\binom{x}{m} := 0$ for each $0 > m \in \mathbf{Z}$ (see §52 [40], [36] and [22]). Therefore,

(8) $(g \circ f)_k = \sum_n g_n \Delta^k \binom{\sum_{m=0}^{\infty} f_m \binom{x}{m}}{n}|_{x=0}$. Since

(9) $\binom{x}{n} = x(x-1)\dots(x-n+1)/n!$, then

(10) $\Delta^k \left(\binom{\binom{f(x)}{m}}{n}\right) = \sum_{k_1+\dots+k_n=k} (k!/(k_1!\dots k_n!)) [\Delta^{k_1} \binom{f(x)}{m}] L^{k_1} [\Delta^{k_2} (\binom{f(x)}{m} - 1)] L^{k_2} \dots [\Delta^{k_{n-1}} (\binom{f(x)}{m} - n + 2)] L^{k_{n-1}} [\Delta^{k_n} (\binom{f(x)}{m} - n + 1)]/n!$, where

(11) $\Delta^k \binom{f(x)}{m} = m^{-1} \sum_{l=0}^{k} \binom{k}{l} [\Delta^l L^{k-l} \binom{f(x)}{m-1} - \delta_{l,0}(m-1)]$. Since

(12) $\Delta^l L^{k-l} f(x)|_{x=0} = \sum_{m=0}^{\infty} f_m \Delta^l \binom{x+k-l}{m}|_{x=0} = \sum_m f_m \binom{k-l}{m-l}$, then

$$(13)\ \Delta^k \left(\binom{\binom{f(x)}{m}}{n}\right)|_{x=0} = (n!)^{-1} \sum_{l_1+\dots+l_n=k} \binom{k}{l_1} \binom{k-l_1}{l_2} \dots \binom{k-l_1-\dots-l_{n-2}}{l_{n-1}}$$

$$[\sum_{m=0}^{\infty} f_m \binom{k-l_1}{m-l_1} - \delta_{l_1,0}(n-1)][\sum_{m=0}^{\infty} f_m \binom{k-l_1-l_2}{m-l_2} - \delta_{l_2,0}(n-2)] \dots$$

$$[\sum_{m=0}^{\infty} f_m \binom{k-l_1-\dots-l_{n-1}}{m-l_{n-1}} - \delta_{l_{n-1},0}]f_{l_n},$$

so that it is necessary to evaluate coefficients

$$(14)\ \Omega^{k,n}_{m_1,\dots,m_n} := \sum_{l_1+\dots+l_{n-1}=k-m_n} \binom{k}{l_1}\binom{k-l_1}{l_2}\dots\binom{k-l_1-\dots-l_{n-2}}{l_{n-1}}\binom{k-l_1}{m_1-l_1}$$

$$\binom{k-l_1-l_2}{m_2-l_2}..\binom{k-l_1-\dots-l_{n-1}}{m_{n-1}-l_{n-1}}.$$

There are identities:

(15) $\binom{m}{l} = (-1)^l\binom{l-m-1}{l}$,

(16) $(1+x)^e(1+(1+x)^a)^b = \sum\binom{b}{c}\binom{ac+e}{d}x^d$, such that for $e=b$ and $a=-1$ this gives

$$(17)\ \sum_{l_1,l_2}\binom{k-l_1}{m_1-l_1}\binom{k-l_1-l_2}{m_2-l_2}\binom{k}{l_1}\binom{k-l_1}{l_2} =$$

$$\sum_{l_1}\binom{k-l_1}{m_1-l_1}\binom{k}{l_1}\binom{k-l_1}{m_2}2^{m_2} =$$

$$\binom{k}{m_1}\sum_{l_1}\binom{m_1}{l_1}\binom{k-l_1}{m_2}2^{m_2},\ \text{since}$$

$$\sum_l \binom{b-l}{m-l}\binom{b}{l} = \sum_l b!/[(m-l)!(b-m+l)!l!].$$

General expressions are too complicated, but it is suffucient to construct a generating function for these coefficients:

$$(18)\ y_1^{m_1}\dots y_{n-2}^{m_{n-2}}y_{n-1}^{m_{n-1}+m_n}(x_1y_2\dots y_n + (y_1+z_2)(x_2y_3\dots y_{n-1}+$$

$$(y_2+z_3)(x_3y_4\dots y_{n-1}+(y_3+z_4)(x_4y_5\dots y_{n-1}+\dots))\dots)^k =$$

$$y_1^{m_1}\dots y_{n-2}^{m_{n-2}}y_{n-1}^{m_{n-1}+m_n}\sum_{l_1}\binom{k}{l_1}(x_1y_2\dots y_{n-1})^{l_1}(y_1+z_2)^{k-l_1}(x_2y_3\dots y_{n-1}+$$

$$(y_2+z_3)(x_3y_4\dots y_{n-1}+(y_3+z_4)(x_4y_5\dots y_{n-1}\dots+(x_{n-1}+y_{n-1})\dots)^{k-l_1} =$$

$$y_2^{m_2}\dots y_{n-2}^{m_{n-2}}y_{n-1}^{m_{n-1}+m_n}\sum_{l_1,q_1,l_2}\binom{k}{l_1}\binom{k-l_1}{q_1-l_1}\binom{k-l_1}{l_2}(x_1y_2\dots y_{n-1})^{l_1}y_1^{k-q_1+m_1}z_2^{q_1-l_1}$$

$$(x_2y_3...y_{n-1})^{l_2}$$

$$(y_2+z_3)^{k-l_1-l_2}(x_3y_3...y_{n-1}+(y_3+...)...)^{k-l_1-l_2}=$$

$$y_3^{m_3}...y_{n-2}^{m_{n-2}}y_{n-1}^{m_{n-1}+m_n}\sum_{l_1,q_1\ l_2,q_2}\binom{k}{l_1}\binom{k-l_1}{q_1-l_1}\binom{k-l_1}{l_2}\binom{k-l_1-l_2}{q_2-l_2}$$

$$(x_1y_3...y_{n-1})^{l_1}y_1^{k+m_1-q_1}z_2^{q_1-l_1}(x_2y_3...y_{n-1})^{l_2}y_2^{k+m_2-q_2}z_3^{q_2-l_2}$$

$$(x_3y_4...y_{n-1}+(y_3+z_4(x_4y_5...y_{n-1}+...)...)^{k-l_1-l_2}=....$$

In this series coefficients in front of $x_1^{m_1}x_2^{m_2}...x_{n-1}^{m_{n-1}}(y_1...y_{n-1})^k$ are equal to $\Omega^{k,n}_{m_1,...,m_n}$, where $x_j=z_{j+1}$ for each $j=1,...,n-1$, x_j and y_j are variables. Therefore, the generating function has the form:

$$(19)\ y_1^{m_1}...y_{n-2}^{m_{n-2}}y_{n-1}^{m_{n-1}-m_n}(x_1y_2...y_n+(x_1+y_1)(x_2y_3...y_{n-1}+(x_2+y_2)(x_3y_4...y_{n-1}+$$

$$(x_3+y_3)(x_4y_5...y_{n-1}...+(x_{n-1}+y_{n-1})...)^k=$$

$$\sum_{m_1,...,m_{n-1}}\Omega^{k,n}_{m_1,...,m_n}x_1^{m_1}x_2^{m_2}...x_{n-1}^{m_{n-1}}(y_1...y_{n-1})^k+$$

$$\sum_{m_1,q_1,...,m_{n-1},q_{n-1}}\Upsilon^{k,n}_{m_1,q_1,...,m_{n-1},q_{n-1}}x_1^{q_1}x_2^{q_2}...x_{n-1}^{q_{n-1}}\times$$

$$y_1^{k+m_1-q_1}...y_{n-2}^{k+m_{n-2}-q_{n-2}}y_{n-1}^{k+m_{n-1}-q_{n-1}+m_n},$$

where coefficients $\Upsilon^{k,n}_{m_1,q_1,...,m_{n-1},q_{n-1}}$ are given by Equation (18).

In particular $id=f^{-1}\circ f$ and $\Delta^k id(x)|_{x=0}=\delta_{k,1}$, hence

(21) $\delta_{k,1}=\sum_{n=0}^{\infty}(n!)^{-1}(f^{-1})_nQ^{k,n}(f)$, where coefficients $Q^{k,n}$ are given by Equations (8, 13, 14, 19), that is

$$(22)\ Q^{k,r}(f)=\sum_{m_1,...,m_n}\ \sum_{i_1<...<i_{n-p};p}\Omega^{k,n}_{m_{i_1},...,m_{i_{n-p}}}f_{m_{i_1}}...f_{m_{i_{n-p}}}$$

$$\prod_{(j_s\in(1,...,n)\setminus\{i_1,...,i_{n-p}\};j_1<j_2<...;l_{j_1}=0,...,l_{j_{p-1}}=0)}(j_s-n)\binom{k-l_1-...-l_{s-1}}{m_s}.$$

For analytic functions there are equalities:

$$(23)\ \Delta^kx^n=\sum_{l_1=0}^{n-1}\binom{n}{l_1}\Delta^{k-1}x^{l_1}=$$

$$\sum_{l_1=0}^{n-1}\sum_{l_2=0}^{l_1-1}\dots\sum_{l_k=0}^{l_{k-1}-1}\binom{n}{l_1}\binom{l_1}{l_2}\dots\binom{l_{k-1}}{l_k}x^{l_k}, \text{ consequently,}$$

$$(24)\ (\Delta^k x^n)_{x=0} = \sum_{l_1=0}^{n-1}\sum_{l_2=0}^{l_1-1}\dots\sum_{l_{k-1}=0}^{l_{k-2}-1}\binom{n}{l_1}\binom{l_1}{l_2}\dots\binom{l_{k-2}}{l_{k-1}} =: T_{n,k}.$$

On the other hand,

(25) $\binom{x}{m} = (m!)^{-1}\sum_{l=0}^{m} x^{m-l}(-1)^l\alpha_l(1,...,m-1) = \sum_l S_{m,l}x^l$, where

(26) $\alpha_l(z_1,...,z_m) := \sum_{i_1<i_2<...<i_l} z_{i_1}...z_{i_l}$, consequently,

(27) $\sum_l S_{m,l}T_{l,j} = \delta_{m,j}$ and

(28) $\sum_l T_{m,l}S_{l,j} = \delta_{m,j}$. Then for analytic g and f:

(29) $g(x) = \sum_m a_m(g)x^m$ and

$$(30)\ (g\circ f)(x) = \sum_{m,l_j,k_j,n} a_m(g)[m!/(l_1!...l_m!)]a_{k_1}^{l_1}(f)...a_{k_m}^{l_m}(f)]x^n,$$

where in the last series $k_1l_1 + ... + k_ml_m = n$, $l_1 + ... + l_m = m$, $0 \le l_j, k_j \in \mathbf{Z}$.

For estimations of $\Omega^{k,n}_{m_1,...,m_n} f_{m_1}...f_{m_n}$ it can be used

(31) $|\binom{k}{l}|_p = p^{-\lambda(k)+\lambda(q)+\lambda(k-q)}$, where

(32) $\lambda(n) = (n-s_n)/(p-1)$, $n = a_0 + a_1p + ... + a_jp^j$, $s_n := a_0 + a_1 + ... + a_j$, where $a_i \in \{0,1,...,p-1\}$, $\lambda(q)+\lambda(k-q)-\lambda(k) = (s_k - s_q - s_{k-q})/(p-1)$ (see also §25 in [40] and [36]).

5. One-Parameter Subgroups of Diffeomorphism Groups

1. Theorem. *If M is a compact manifold over a locally compact field $\mathbf{K}$ of characteristic $char(\mathbf{K}) = p > 1$, $1 \le t \in \mathbf{N}$, $\alpha \in \{t,[t]\}$, then there exists a clopen subgroup W in $Diff^\alpha(M)$ such that each element $g \in W$ lies on a local one-parameter subgroup g^x continuous by x relative to the multiplicative group $(\mathbf{K}^*,\times)$ with $x \in \mathbf{K}^*$.*

Proof. Since M is compact, then M is finite dimensional $dim_{\mathbf{K}}M = n$ over $\mathbf{K}$ and the manifold M can be supplied with a finite disjoint analytic atlas, that is, with disjoint clopen charts a finite union of which covers M. Therefore, there exists a natural C^α embedding of M into $\mathbf{K}^n$ as the clopen subset. As it was proved above $Diff^\alpha(M)$ is mertizable with the left-invariant metric ρ^α. Choose

$k_0 \in \mathbf{N}$ and the clopen subgroup $W := \{g \in Diff^\alpha(M) : \rho^\alpha(g, id) < |\pi|^{k_0}\}$ such that $g \in W$ implies $\|g - id\|_{C^\alpha_b(M,\mathbf{K}^n)} < |\pi|^{k_0}$.

Consider $x \in B(\mathbf{K}, 1, |\pi|^s)$, where $1 \le s \in \mathbf{N}$, then $|x| = 1$, since $|x-1| \le |\pi|^s$ and due to the ultrametric inequality. If $g \in W$, then for the proving of this theorem it is necessary to find a one-parameter local subgroup $\{g^x : x \in B(\mathbf{K}, 1, |\pi|^s)\}$ satisfying Conditions $(1-3)$:

(1) $g^1 = id$;

(2) $g^{x_1} g^{x_2} = g^{x_1 x_2}$ for each $x_1, x_2 \in B(\mathbf{K}, 1, |\pi|^s)$;

(3) $g^{x_0} = g$ for some $x_0 \in B(\mathbf{K}, 1, |\pi|^s)$.

Each $g \in Diff^\alpha(M)$ has the form $g = (g_1, ..., g_n)$, where $g_j : M \to \mathbf{K}$ for each $j \in \{1, ..., n\}$. Since M is embedded as the compact subset in $\mathbf{K}^n$, then its covering by balls $B(\mathbf{K}^n, x, r) \subset M$ has a finite subcovering $\{B(\mathbf{K}^n, x_j, r_j) : j = 1, ..., m\}$, where $x \in M$, $0 < r_j < \infty$, $m \in \mathbf{N}$, consequently, $\min_{j=1,...,m} r_j = r > 0$.

If g^{x_1} and f^{x_2} are two commuting local one-parameter subgroups for each $x_1, x_2 \in B(\mathbf{K}, 1, |\pi|^s)$ in $Diff^\alpha(M)$, then $g^x f^x =: (gf)^x$ is a local one-parameter subgroup in $Diff^\alpha(M)$, since $(gf)^{x_1}(gf)^{x_2} = (g^{x_1} f^{x_1})(g^{x_2} f^{x_2}) = g^{x_1} g^{x^2} f^{x_1} f^{x_2} = g^{x_1 x_2} f^{x_1 x_2} = (gf)^{x_1 x_2}$ for each $x_1, x_2 \in B(\mathbf{K}, 1, |\pi|^s)$, also $g^1 = f^1 = e = id$. For example, if $supp(g^x) \subset A$ and $supp(f^x) \subset C$ for each $x \in B(\mathbf{K}, 1, |\pi|^s)$ such that $A \cap C = \emptyset$, where A and C are closed subsets in M, $supp(g) := cl_M\{y \in M : g(y) \ne y\}$, $cl_M(V)$ denotes the closure of a subset V in M, then $g^{x_1} f^{x_2} = f^{x_2} g^{x_1}$ for each $x_1 \ne x_2 \in B(\mathbf{K}, 1, |\pi|^s)$. On the other hand, each $g \in Diff^\alpha(M)$ can be decomposed into the product $g = h_1 ... h_m$, where $supp(h_j) \subset B(\mathbf{K}^n, x_j, r_j)$ for each $j = 1, ..., m$. Therefore, the proof of this theorem reduces to the case $Diff^\alpha(B)$, where B is a clopen compact ball in $\mathbf{K}^n$, since W decomposes into the internal direct product of its subgroups $W_j := \{g \in W : supp(g) \subset B(\mathbf{K}^n, x_j, r_j)\}$ and W_j has the natural embedding into $Diff^\alpha(B(\mathbf{K}^n, x_j, r_j))$ for each $j = 1, ..., m$. For $g = id$ put $id^x = id$ for each $x \in B(\mathbf{K}, 1, |\pi|^s)$. Therefore, due to Equations 3.3$(1,2)$ it remains the case of $g \ne id$ in W, consequently, $x_0 \ne 1$ for such g.

In accordance with Lemma 7.6 [15] if $\mathbf{K}$ is a field and G is a finite subgroup of the multiplicative group of nonzero elements of $\mathbf{K}$, then G is a cyclic group. Then by Theorem 7.b the multiplicative group of nonzero elements of a finite field is cyclic. In view of the Wedderburn Theorem 7.c a finite division ring is necessarily a commutative field [15]. The field $\mathbf{K}$ is isomorphic with $\mathbf{F}_{p^u}(\theta)$, where $\mathbf{F}_{p^u}$ is the finite field consisting of p^u elements, $u \in \mathbf{N}$. The finite field $\mathbf{F}_{p^u}$ is the splitting field of the polynomial $x^{p^u} - x$. Thus $\chi : \mathbf{K}^* \to \mathbf{K}^*$ is a continuous multiplicative character, where $\mathbf{K}^* = \mathbf{K} \setminus \{0\}$, if and only if it can be written in the form $\chi(x) = \phi(x)\psi(x)$, where ϕ and ψ are continuous multiplicative characters such that $\phi : \mathbf{F}^*_{p^u} \to \mathbf{F}^*_{p^u}$ is some homomorphism of the multiplicative group of $\mathbf{F}_{p^u}$ and $\phi(\theta) = \theta$, $\psi(\theta) = \theta^k$ for some nonnegative integer k and the restriction of ψ on $\mathbf{F}_{p^u}$ is the identity mapping, since there exists the natural embedding of $\mathbf{F}_{p^u}$ into $\mathbf{K}$. Therefore, $|\chi(\theta)| = |\theta|^k$. Denote by Ω the family of all continuous multiplicative characters $\chi : \mathbf{K}^* \to \mathbf{K}^*$. In particular, $\chi_1(x) = 1$ and $\chi_{id}(x) = x$ for each in $x \in \mathbf{K}^*$ are the trivial character and the identity character respectively.

If g^x is a local one-parameter subgroup in $Diff^t(M)$, then $g^{\chi(x)}$ is also a local one-parameter subgroup in $Diff^{\alpha}(M)$. If $g^{x_0} = g$, then $g^z = g$ for each $\chi(z) = x_0$ with $\chi \in \Omega$, since $\chi(1) = 1$. Therefore, if $x_0 \in B(\mathbf{K}, 1, |\pi|^s)$ is a marked point, then it is sufficient to satisfy Condition (3) for some $x \in V_s := \{\chi^{-1}(x_0) : \chi \in \Omega\} \cap B(\mathbf{K}, 1, |\pi|^s)$, where $x_0 \neq 1$ and $x \neq 1$ for $g \neq 1$. Thus, if each $h_j \neq id$ in W_j has $x_0(h_j) \in V_s$, then the desired local one-parameter subgroup g^x will be found for $id \neq g = h_1 \dots h_m \in W$. Take, for example, $x_0 = 1 + \theta^s$.

We say that a family A of functions $f : X \to \mathbf{K}$ separate points of X if for each $x \neq y \in X$ there exists $f \in A$ such that $f(x) \neq f(y)$. If G is a cyclic group of order k and $a \in G$ is such that $\{1, a, a^2, \dots, a^{k-1}\} = G$ and m is a natural number mutually prime with k, $(m,k) = 1$, then $\zeta : G \to G$ such that $\zeta(a^l) = a^{lm}$ for each $l = 0, \dots, k-1$ is the automorphism of G, in particular, for $k = p^u - 1$. Since the multiplicative group $G = \mathbf{F}^*_{p^u}$ is cyclic, then the family of all ϕ separate points of G, hence the family Ω of all continuous multiplicative characters χ of $\mathbf{K}^*$ separate points of $\mathbf{K}^*$.

In view of the Kaplansky theorem a subalgebra A of $C^0(X, \mathbf{K})$ containing all constant functions and separating points of a locally compact totally disconnected Hausdorff space X is dense in $C^0(X, \mathbf{K})$ [18, 3]. From [1] it follows, that $C^0(M, \mathbf{K}^n)$ has the polynomial basis $\{Q_{\bar{m}}(y)e_i : \bar{m} \in \mathbf{N_0}^n, i = 1, \dots, n\}$ with $N_0 = \{0, 1, 2, \dots\}$ and $e_i = (0, \dots, 0, 1, 0, \dots) \in \mathbf{K}^n$ with 1 on the i-th place.

In view of the proof above up to the affine C^∞ diffeomorphism it is sufficient to consider W for $Diff^t(B)$, where $B = B(\mathbf{K}^n, 0, 1)$. We seek a solution in the form

(4) $g^x_j(y) = y_j + \sum_{k=1}^{\infty} b_{j,k}(y, x)$

for each $j = 1, \dots, n$ with the converging series of functions $b_{j,k}(y,x) : M \times B(\mathbf{K}, 1, |\pi|^s) \to \mathbf{K}$ of class C^t by $y \in M$ and continuous by $x \in B(\mathbf{K}, 1, |\pi|^s)$ for each $j = 1, \dots, n$ such that $\|b_{j,k}\|_{C^t_b(B,\mathbf{K})} < |\pi|^{k_0}$ and $\lim_{k\to\infty} \max_{j=1,\dots,n} \|b_{j,k}\|_{C^t(B,\mathbf{K})} = 0$, where $g = (g_1, \dots, g_n)$. Each function $b_{j,k}(y,x)$ has the decomposition

(5) $b_{j,k}(y,x) = \chi_{j,k}(x) \sum_{\bar{m} \in \mathbf{N_0}^n} c_{j,k,\bar{m}} Q_{\bar{m}}(y)$,
where $\chi_{j,k}(x) \in \Omega$, $y \in B$, $x \in B(\mathbf{K}, 1, |\pi|^s)$, $c_{j,k,\bar{m}} \in \mathbf{K}$. Without loss of generality we can suppose that $|\chi_{j,k}(x)| \le 1$ on $B(\mathbf{K}, 1, |\pi|^s)$ for each j,k,m, since B is the unit ball and $s \ge 1$. The convergence of Series (4) is equivalent to:

(6) $\lim_{k+|\bar{m}| \to \infty} \max_{j=1,\dots n} \|c_{j,k,\bar{m}} Q_{\bar{m}}(y)\|_{C_b^{\alpha}(M,\mathbf{K})} \|\chi_{j,k}(x)\|_{C_b^0(B(\mathbf{K},1,|\pi|^s),\mathbf{K})} =$
0,
where $|\bar{m}| := m_1 + \dots + m_n$, $m = (m_1, \dots, m_n)$. Then Condition (1) is equivalent to $\sum_{k=1}^{\infty} c_{j,k,\bar{m}} = 0$ for each $\bar{m}$ and j, while Condition (3) is equivalent to $\sum_{\bar{m}} c_{j,\bar{m}} Q_{\bar{m}}(y) = g_j(y) - y_j$ with $c_{j,\bar{m}} := \sum_{k=1}^{\infty} c_{j,k,\bar{m}} \chi_{j,k}(x_0)$, since $\chi_{j,k}(1) = 1$, $\{Q_{\bar{m}}(y) : \bar{m} \in \mathbf{N_0}^n\}$ are linearly independent over $\mathbf{K}$.

Consider two monotone increasing sequences $\{s_v \in \mathbf{N} : v \in \mathbf{N}\}$ and $\{q_v \in \mathbf{N} : v \in \mathbf{N}\}$ such that $q_v \ge s_v$ for each v. For each $q > l \in \mathbf{N}$ there exists the quotient algebraic homomorphism of rings $\pi_l^q : B(\mathbf{K}, 0, 1)/B(\mathbf{K}, 0, |\pi|^q) \to B(\mathbf{K}, 0, 1)/B(\mathbf{K}, 0, |\pi|^l)$ and $\pi_q : B(\mathbf{K}, 0, 1) \to B(\mathbf{K}, 0, 1)/B(\mathbf{K}, 0, |\pi|^q)$ such that $\pi_l^q \circ \pi_q = \pi_l$. These algebraic homomorphisms induce natural homomorphisms of functions in the same notation such that $\pi_q^*(g) =: {}_q g$, where ${}_q g \in Hom(M_q)$, $M_q = \pi_q(M)$ is a finite set, $g \in W \subset Diff^t(B)$, ${}_q g : M_q \to M_q$ is a bijective surjective mapping of M_q onto M_q (see Formulas 3.2$(4-7)$). Then also $\pi_{s_v}(B(\mathbf{K}^*, 1, |\pi|^s))$ is the finite set for $s_v > s$. For $s_v > 1$ we have that $\pi_{s_v}(\mathbf{K})$ is the discrete commutative ring (see §3.1). If $x \in \mathbf{K}$ and $|x| = 1$, then $|x^{-1}| = 1$. The multiplicative norm in $\mathbf{K}$ induces the norm in $\pi_{s_v}(\mathbf{K})$. If $|x-1| < 1$, then $x^{-1} = (1 + (x-1))^{-1} = 1 + \sum_{l=1}^{\infty} (1-x)^l$, thus $\pi_{s_v}(x)$ has the inverse in $\pi_{s_v}(B(\mathbf{K}, 1, |\pi|^s))$, when $|x-1| \le |\pi|^s$, $s \ge 1$, since $\pi_{s_v}((x-1)^l) = 0$ for $ls \ge s_v$. Then $1 \in \pi_{s_v}(B(\mathbf{K}, 1, |\pi|^s)$ and $\pi_{s_v}(x_1 x_2) = \pi_{s_v}(x_1) \pi_{s_v}(x_2) \in \pi_{s_v}(B(\mathbf{K}, 1, |\pi|^s))$ for each $x_1, x_2 \in B(\mathbf{K}, 1, |\pi|^s)$, since $|x_1 x_2 - 1| \le \max(|x_1 - 1||x_2|, |x_2 - 1|) \le |\pi|^s$ and $|x_2| = 1$. Thus, $B(\mathbf{F}_{p^u}, 1, |\pi|^s)$ for a natural number $s \ge 1$ is the multiplicative commutative group and $B(\mathbf{F}_{p^u}, 1, |\pi|^{s_1})/B(\mathbf{F}_{p^u}, 1, |\pi|^{s_2})$ for each natural numbers $1 \le s_1 < s_2$ is the finite multiplicative commutative quotient group.

If $g \in W$ and $k_0 \ge 1$, then $|g(y) - y| \le \|g - id\|_{C^1} |y| \le |\pi|^{k_0} |y|$ and $|g(y_1) - g(y_2)| \le \|g\|_{C^1} |y_1 - y_2|$ and $|g^{-1}(y_1) - g^{-1}(y_2)| \le \|g^{-1}\|_{C^1} |y_1 - y_2|$, since $t \ge 1$, consequently, g is the isometry, that is, $|g(y)| = |y|$ for each $y \in M$.

For given s and s_1 choose k_0 and $q_1 \ge s_1$ such that for each $g \in W$ we have ${}_{q_1} g = id$ on M_{q_1}. Then for a given $g \in W$ there exists the algebraic homomorphism η_1 from $\pi_{s_1}(B(\mathbf{K}, 1, |\pi|^s))$ into $\pi_{q_1}(W)$ satisfying Conditions $(1-3)$ with $\pi_{s_1}(x_0)$ instead of x_0. Each $g \in W$ and each $x \in \mathbf{K}$ is the projective limit of the inverse sequences $g = pr - \lim\{ {}_{q_v} g, \pi_{q_l}^{q_v}, \Lambda_{q_1}\}$ and $x = pr - \lim\{ {}_{s_v} x, \pi_{s_l}^{s_v}, \mathbf{N}\}$, since sequences $\{q_v\}$ and $\{s_v\}$ are cofinal with $\mathbf{N}$, where $v \ge l \in \mathbf{N}$ (see also §2.5

[11]). For example, we can take $s_{v+1} = s_v + 1$ for each $v \in \mathbf{N}$. Consider ${}_{q_{v+1}}g$, then $\pi_{q_v}^{q_{v+1}}({}_{q_{v+1}}g) = {}_{q_v}g$. Each ${}_{q_v}g \in Hom(M_{q_v})$ is the element of the symmetric group S_{b_v}, where b_v is the cardinality of the finite set $\pi_{q_v}(M) = M_{q_v}$ (see also Theorem 3.6 above). Thus, ${}_{q_v}g$ has a decomposition into a product of nonintersecting finite cycles. If $y_v \in M_{q_v}$, then the cardinality of each $(\pi_{q_v}^{q_{v+1}})^{-1}(y_v)$ is the same for each $y_v \in M_{q_v}$. The homomorphism ${}_{q_{v+1}}g$ on each subset $(\pi_{q_v}^{q_{v+1}})^{-1}(y_v)$ acts as the isometry. Then each cycle of ${}_{q_v}g$ splits into the product of cycles or becomes a cycle of greater length for ${}_{q_{v+1}}g$. Moreover, ${}_{q_{v+1}}g$ is the product of two homomorphisms h and f, where $f((\pi_{q_v}^{q_{v+1}})^{-1}(y_v)) = (\pi_{q_v}^{q_{v+1}})^{-1}(y_v)$ for each $y_v \in M_{q_v}$ and $\pi_{q_v}^{q_{v+1}}(f) = \pi_{q_v}^*(id)$ is the identity on $\pi_{q_v}(M)$, while $h(z_{v+1}) = h(y_{v+1,0}) + (z_{v+1} - y_{v+1,0})$ with a marked $y_{v+1,0} \in (\pi_{q_v}^{q_{v+1}})^{-1}(y_v)$ and for each $z_{v+1} \in (\pi_{q_v}^{q_{v+1}})^{-1}(y_v)$.

If $\sigma \in \Sigma_l$ and σ is a cycle of length u, then its algebraic order is u, that is, the cyclic group $\{\sigma^a : a \in \mathbf{N}\}$ is of order u. On the step v there is the commutative local subgroup ${}_{q_v}g^{x_v} = \eta_v(x_v)$ for each $x_v \in \pi_{s_v}(B(\mathbf{K}, 1, |\pi|^s))$. Therefore, on the $v+1$-th step choose q_{v+1} sufficiently large such that a number of nonintersecting cycles $\sigma_{v,j}$ or their length are sufficiently large in the sense that a commutative subgroup of $\pi_{q_{v+1}}(W)$ generated by finite products of $\sigma_{v,j}^{b_j}$, $b_j \in \mathbf{N}$, is of sufficiently large order that to provide the algebraic homomorphism η_{v+1} from $\pi_{s_{v+1}}(B(\mathbf{K}, 1, |\pi|^s))$ into $\pi_{q_{v+1}}(W)$ satisfying Conditions $(1-3)$ with $\pi_{q_{v+1}}(x_0)$ instead of x_0 and $\pi_{s_v}^{s_{v+1}}(\eta_{v+1}) = \eta_v$.

In accordance with Lemma 2.5.10 [11] if $\{\phi, f_{j'}\}$ is a mapping of an inverse system $S = \{X_j, \pi_i^j, \Psi\}$ into an inverse system $S' = \{Y_{j'}, \pi_{i'}^{j'}, \Psi'\}$ and all $f_{j'}$ are homeomorphisms, then the limit mapping $f = pr - \lim\{\phi, f_{j'}\}$ is also the homeomorphism of $X = pr - \lim S$ onto $Y = pr - \lim S'$, where $\phi : \Psi' \to \Psi$ is a nondecreasing function, Ψ' and Ψ are directed sets.

Using the same notation which can not cause a confusion we mention the following. Each this induction step gives the corresponding solution of $(4-6)$ with $\eta_v(\pi_{s_v}(x)) \in \pi_{q_v}(B(\mathbf{K}, 1, |\pi|^s))$ and $\pi_{q_v}(y)$ and $\pi_{q_v}(c_{j,k,\bar{m}})$, $\pi_{q_v}^*(Q_{\bar{m}}(y))$ and $\pi_{s_v}(\chi_{j,k}(x)) \in \pi_{q_v}(B(\mathbf{K}, 1, |\pi|^s))$ instead of $x, y, c_{j,k,\bar{m}}$, $Q_{\bar{m}}(y)$ and $\chi_{j,k}(x)$ respectively. If $f \in C_b^\alpha(M, \mathbf{K}^n)$ and $\|f - id\|_{C_b^\alpha(M,\mathbf{K}^n)} < 1$, then $f \in Diff^\alpha(M)$. We say that a subset A in M is an ε-net, if for each $y \in M$ there exists $z \in A$ such that $|z - y| < \varepsilon$, where $0 < \varepsilon < \infty$. We have $\lim_{v\to\infty} |\pi|^v = 0$ and representatives of M_{q_v} in M form a ρ^v net for suitable subsequence q_v, which is denoted by the same notation, where $0 < \rho < 1$ is a constant and ρ is a parameter characterizing a net A_{q_v} corresponding to $\{Q_{\bar{m}} : \bar{m} \in \mathbf{N_0}^n\}$ for $m_j \le q_v$ for each

$j = 1, ..., n$, where each $z \in A_{\bar{c}_v}$ is a zero of $Q_{\bar{m}}$ as soon as $m_j \geq m_{j,0}$ for some $j = 1, ..., n$, where $\bar{m}_0 := (m_{1,0}, ..., m_{n,0})$ corresponds to z. On the other hand, $|\chi_{j,k,m}(x)| \leq 1$ on $B(\mathbf{K}, 1, |\pi|^s)$ for each j, k, m, moreover, $x_0 = 1 + \theta^s$, $|x_0| = 1$. Then the corresponding to it series converges, since ${}_{q_v}g^{\eta_v(x_v)} \in \pi^*_{q_v}(W)$ for each $x_v \in \pi_{s_v}(B(\mathbf{K}, 1, |\pi|^s))$ and W is complete relative to the C^α_b uniformity. Thus, the inverse sequence of ${}_{q_v}g^{\eta_v(\pi_{s_v}(x))}$ converges to a local multiplicative continuous one-parameter subgroup $g^x(y) \in W$ relative to the $C^\alpha_b \times C^0_b$ uniformity which is C^α_b by $y \in M$ and C^0_b by $x \in B(\mathbf{K}, 1, |\pi|^s)$.

2. Theorem. *If M is a manifold on the Banach space $X := c_0(\gamma_X, \mathbf{K})$ with a finite atlas and charts with bounded $\phi_j(U_j)$ in X for each $j \in \Lambda_M$, where $char(\mathbf{K}) = p > 1$, then in $Diff^\alpha_b(M)$ in each neighborhood of id there are $g \neq id$ which does not belong to any non-trivial one-parameter subgroup g^y relative to the additive group $(\mathbf{K}, +)$. Moreover, if M is embedded into X and*

(i) $(h - h \circ g^{-1})$, $(h - h \circ g^{-1}) \circ g^2, ..., (h - h \circ g^{-1}) \circ g^{p-1}$ are $\mathbf{K}$-linearly independent and $g \neq id$, where $h := g - id$, then g does not belong to any one-parameter subgroup $\{g^y : y \in (\mathbf{K}, +)\}$.

Proof. From the conditions of this theorem it follows, that M has a C^t embedding as an open bounded subset in X. In view of Theorem 2.9 the group $Diff^\alpha_b(M)$ is metrizable. Since $Diff^\beta_b(M)$ is everywhere dense in $Diff^\alpha_b(M)$ for each $\beta > \alpha$, then it is sufficient to consider the case $t \geq 1$, $\alpha \in \{t, [t]\}$. If $g \in Diff^\alpha_b(M)$, then up to a diffeomorphism of manifolds $g|_U \in Diff^\alpha_b(U)$ for any U open in M, since $g : U \to g(U) \subset M$. Let $U \subset M$ be a clopen bounded subset such that $U \subset U_j$ for some $j \in \Lambda_M$, $g \in Diff^\alpha_b(M)$ and $supp(g) \subset U$, then $g|_U \in Diff^\alpha_b(U)$. Take in particular $U := \phi_j^{-1}(B(X, 0, |\pi|))$, where $\pi \in \mathbf{K}$, $0 < |\pi| < 1$. Thus, if prove theorem for $U = B(X, 0, |\pi|)$, then it will be also true for $Diff^\alpha_b(M)$.

Consider in $Diff^\alpha_b(U)$ a left-invariant metric ρ^α (see Theorem 2.9). Take $id \neq g \in W := \{f \in Diff^\alpha_b(U) : \rho^\alpha(f, e) \leq |\pi|\}$, then $g = id + h$, where $0 < \|h\|_{C^\alpha_b(U,X)} \leq |\pi|$. If we consider p^n as the element of $\mathbf{K}$, where $n \in \mathbf{N} := \{1, 2, ...\}$, then $p^n = 0 \in \mathbf{K}$, since $char(\mathbf{K}) = p$. So we need to have $g^{p^n} = g^0 = id$ on U, where $g^1 = g$, $g^{k+1} = g^k g^1, ..., g^{p^n} = g^{p^n - 1} g^1$. Consider $(1 + a)^k = \sum_{s=0}^{l} \binom{k}{s} a^s$, where $a \in \mathbf{K}$, then $(1 + a)^{p^n} = 1 + a^{p^n}$ and inevitably $|(1 + a)^{p^n}| = |1 + a^{p^n}|$, since each binomial coefficient $\binom{p^n}{l}$ is divisible on p for each $1 \leq l \leq p^n - 1$ and hence $\binom{p^n}{l}$ is equal to zero in $\mathbf{K}$. Then $(x_i + a x_i^l) \circ (x_i + a x_i^l) = x_i + a x_i^l + a \sum_{s=0}^{l} \binom{l}{s} x_i^{l + (l-1)s} a^s$ and so on. In particular,

$(x_i+ax_i^p)\circ(x_i+ax_i^p)=x_i+2ax_i^p+x_i^{p^2}a^{p+1}$ for $l=p$ and so on. This shows, that elements of the form $g(x)=id(x)+ax_i^le_i$ with $a\neq 0$ and $1<l\in\mathbf{N}$ can not lie on any one-parameter subgroups, since $g^p\neq id$.

Demonstrate in general for M embedded into X, that each $id\neq g\in W$ satisfying Condition (i) does not belong to any one-parameter subgroup. On the other hand, $g^2(x)=g\circ g(x)=id\circ(id+h)(x)+h\circ(id+h)(x)=g(x)+h\circ g(x)$, $g^n(x)=g^{n-1}(x)+h\circ g^{n-1}(x)$, consequently, $\rho^\alpha(g^n,g^{n-1})=\rho^\alpha(g^2,g)\leq|\pi|$ for each $n\in\mathbf{N}$ and $\|h\circ g^{n-1}\|_{C_b^\alpha}\leq|\pi|$. Then by induction $g^n(x)=g(x)+h\circ g(x)+...+h\circ g^{n-1}(x)$ for each $n\geq 2$, hence $\rho^\alpha(g^n,g)\leq|\pi|$. Then $|h(y)-h(x)|\leq\|h\|_{C_b^\alpha(U,X)}|x-y|$ for each $x,y\in U$, in particular, for $y=g(x)$, where $y-x=g(x)-x=h(x)$, hence $|h\circ g(x)-h(x)|\leq\|h\|_{C_b^\alpha(U,X)}|h(x)|\leq|\pi|^2$. Therefore, $|g^n(x)-g(x)-(n-1)h(x)|\leq\|h\|_{C_b^\alpha(U,X)}|h(x)|$ for each $x\in U$, where $|h(x)|\leq\|h\|_{C_b^\alpha(U,X)}|x|\leq|\pi||x|$, consequently, $|g^n(x)-g(x)-(n-1)h(x)|\leq(\|h\|_{C_b^\alpha(U,X)})^2|x|\leq|\pi|^2|x|$. Thus $|g^{p^k}(x)-id(x)|\leq(\|h\|_{C_b^\alpha(U,X)})^2|x|$, since $|p^kh(x)|=0$. Suppose that $g^{p^k}=id$ for each $k\in\mathbf{N}$, then $0=g^p(x)-id(x)=g^p(x)-g(x)-(p-1)h(x)=h(x)+h\circ g(x)+...+h\circ g^{p-1}(x)=(h\circ g(x)-h(x))+...+(h\circ g^{p-1}(x)-h(x))=h(x)+p(h\circ g^1(x)-h(x))+(p-1)(h\circ g^2(x)-h\circ g(x))+...+3(h\circ g^{p-2}(x)-h\circ g^{p-3}(x))+2(h\circ g^{p-1}(x)-h\circ g^{p-2}(x))-h\circ g^{p-1}(x)$, since $ph(x)=0$ and $p(h\circ g-h)(x)=0$ identically and $g^{p-1}(x)=g^{-1}(x)$. Therefore, it would be $0=|h(x)-h\circ g^{-1}(x)|$ on U, since Condition (i) is supposed to be satisfied, that leads to the contradiction, since $h\neq 0$, $g\neq id$, $h\circ g\neq h$. Thus $g^p(x)\neq id(x)$ for each $id\neq g\in W$ satisfying Condition (i), consequently, g does not belong to any one-parameter subgroup $\{g^y:y\in(\mathbf{K},+)\}$.

2.1. Remark. On the other hand, if $M\supset B(\mathbf{K}^n,x_0,r)$, then $g^y(x):=x+yz$ for $x\in B(\mathbf{K}^n,x_0,r)$ and $g^y(x):=x$ for $x\in M\setminus B(\mathbf{K}^n,x_0,r)$ is the non-trivial one-parameter subgroup for a marked $z\in B(\mathbf{K}^n,0,r)$, where $y\in B(\mathbf{K},0,1)$, $0<r$, $x_0\in M$. Therefore, for each neighborhood W of id there exists $0\neq z\in B(\mathbf{K}^n,0,r)$ such that $id\neq g\in W$.

6. Topological Perfectness of Diffeomorphism Groups

1. Definition. A group G is called algebraically perfect, if its commutator group $[G,G]$ coincides with G, where $[G,G]$ is the minimal group generated by all commutators $[f,g]:=f^{-1}g^{-1}fg$, $f,g\in G$. A group G is called algebraically simple, if it does not contain any normal subgroup other than $\{e\}$ and G. A topological group G is called topologically perfect if $cl_G[G,G]=G$, G is called topologically simple if it does not contain any closed normal subgroup other than $\{e\}$ or G, where $cl_G V$ denotes the closure of a subset V in G.

2. Remark. It is well known that the symmetric group S_n is perfect for $n\neq 2$ and $n\neq 6$, but it is not simple for $n\geq 3$, since the subgroup A_n consisting of even permutations is its normal subgroup different from $\{e\}$ and S_n [19]. Below the topological perfectness and simplicity of the diffeomorphism group $Diff^{\alpha}(M)$ is considered relative to its C^{α} topology.

3. Theorem. *Let M be a compact manifold, $t\geq 0$, $\alpha\in\{t,[t]\}$. Then $Diff^{\alpha}(M)$ supplied with the C^{α} compact-open topology is topologically perfect.*

Proof. In view of Theorem 3.6 above and Theorem 2.5 [24] the diffeomorphism group $Diff^{\alpha}(M)$ algebraically is the projective limit of an inverse sequence $S=\{G_q,\pi^q_s,\mathbf{N}\}$ of finite groups G_q, where G_q is isomorphic with the symmetric group S_b with $b=card(M_q)$, $M_q=\pi_q(M)$. There exists $q_0\in\mathbf{N}$ such that for each $q>q_0$ the cardinality of M_q is greater than 6, consequently, G_q is perfect for each $q>q_0$, since S_b is perfect with $b=card(M_q)$ by the Hölder Theorem 5.3.1 [19]. Each element $h\in Diff^t(M)$ is decomposable into the thread $\{h_q,\pi^q_s,\mathbf{N}\}$. Every h_q is decomposable as a finite product of commutators in G_q for $q>q_0$. Therefore, the commutator group $[G,G]$ is dense in G (see also Theorem 3.7 above), since the projective limit of a thread of products of commutators is the product of commutators, where $G=Diff^{\alpha}(M)$.

4. Corollary. *Let M be a locally compact manifold, $t\geq 0$, $\alpha\in\{t,[t]\}$. Then the diffeomorphism group $Diff^{\alpha}(M)$ and the group $Diff^{\alpha}_c(M)$ of all C^{α} diffeomorphisms with compact supports supplied with the C^{α} compact-open topology are topologically simple.*

Proof. The group $Diff^{\alpha}_c(M)$ is everywhere dense in $Diff^{\alpha}(M)$. Therefore, it is sufficient to prove this corollary for $Diff^{\alpha}_c(M)$.

The group $G=Diff^{\alpha}_c(M)$ satisfies the Epstein system of axioms. Let X be a paracompact Hausdorff topological space, G a group of homeomorphisms of X, and $\mathcal{U}$ a basis of open sets for the topology of X. The Epstein axioms are the following:

$(E1)$ if $U\in\mathcal{U}$ and $g\in G$, then $gU\in\mathcal{U}$;

$(E2)$ G acts transitively on $\mathcal{U}$;

$(E3)$ let $g \in G$, $U \in \mathcal{U}$ and $\mathcal{B}$ be an open cover of X, then there exist $m \in \mathbf{N}$ and $g_1, ..., g_m \in G$ and $V_1, ..., V_m \in \mathcal{B}$ such that:

(i) $g = g_m g_{m-1} ... g_1$;

(ii) $supp(g_i) \subset V_i$;

(iii) $supp(g_i) \cup (g_{i-1} ... g_1 cl(U)) \neq X$ for each $i : 1 \leq i \leq n$.

The manifold M has an C^α embedding θ as the clopen subset into either $\mathbf{K}^n$ or a direct topological sum of copies of $\mathbf{K}^n$ such that $\theta(M)$ is a disjoint union of clopen balls, since M is totally disconnected and has a base of topology consisting of clopen compact subsets in M. As $\mathcal{U}$ take a system of all clopen proper subsets U in M for which there exists $g \in G$ such that $\theta(gU)$ is a ball of finite positive radius in $\mathbf{K}^n$. Since M is totally disconnected and modelled on $\mathbf{K}^n$ as the manifold, then $\mathcal{U}$ forms the base of the topology of M.

In particular, for the diffeomorphism group $G = Diff^\alpha_c(M)$ and the family $\mathcal{U}$ of $M = X$ we have $gU \in \mathcal{U}$ for each $g \in G$, since $g^{-1}(gU) = U$ and $g : M \to M$ is the continuous bijective epimorphism together with g^{-1}. If $U = M$, then $g(M) = M$. A disjoint clopen covering $\mathcal{C}$ of $\theta(M)$ by balls form a C^∞ atlas (moreover, it is the analytic atlas) such that if $f|_U$ is of C^α class for each $U \in \mathcal{C}$, then f is C^α on $\theta(M)$. If $U \in \mathcal{U}$ and $U \neq M$, then $M \setminus U$ is a clopen nonvoid subset in M. Therefore, if U_1 and U_2 are two clopen nonvoid subsets in $\mathcal{U}$ different from M, then there exists $g \in G$ such that $g(U_1) = U_2$, since each two clopen balls in $\mathbf{K}^n$ are analytically diffeomorphic. Thus axioms $(E1, E2)$ are satisfied.

If $g \in G$, then $supp(g) := cl\{x \in M : g(x) \neq x\}$ is compact and for an open covering $\mathcal{B}$ of M there are $V_1, ..., V_m$ in $\mathcal{B}$ such that $supp(g) \subset V_1 \cup ... \cup V_m$. Suppose that $U \in \mathcal{U}$ and $cl(U) \neq M$, where $cl(U)$ denotes the closure of U in M. Since U is clopen in M, then $cl(U) = U$. The topological Tychonoff space M is totally disconnected and locally compact paracompact, since it is modelled on $\mathbf{K}^n$. In view of Theorem 6.2.9 [11] it is strongly zero dimensional. Therefore, $V_1, ..., V_m$ has a refinement $P_1, ..., P_m$ consisting of clopen compact subsets in M such that $supp(g) \subset P_1 \cup ... \cup P_m$. Take $P_0 := \emptyset$ and $W_i := P_i \setminus_{j<i} P_j$, then $W_1, ..., W_m$ is the disjoint clopen covering of $supp(g)$, where $W_i \subset P_i \subset V_i$ for each $i = 1, ..., m$. Therefore, $g(W_1),...,g(W_m)$ is the clopen disjoint covering of $g(supp(g))$. Each $\theta(W_i)$ is a finite union of clopen balls $B_{i,j}$ in $\mathbf{K}^n$. Then put $g_m|_{W_m} = g|_{W_m}$ and $g_m|_{M \setminus W_m} = id$, $g_{m-1}|_{W_{m-1}} = g_m^{-1} g|_{W_{m-1}}$ and $g_{m-1}|_{M \setminus W_{m-1}} = id$, $g_i|_{W_i} = g_{i+1}^{-1} ... g_m^{-1} g|_{W_i}$ and $g_i|_{M \setminus W_i} = id$ for each $i = 1, ..., m-2$, hence Conditions (i, ii) of $(E3)$ are satisfied. For each $1 \leq k < j < m$ there is the identity $g_j ... g_k|_{W_k} = g_j ... g_{k+1} g_{k+1}^{-1} ... g_m^{-1} g|_{W_k} = g_{j+1}^{-1} ... g_m^{-1} g|_{W_k}$ and $g_m ... g_k|_{W_k} = g|_{W_k}$ for each $1 \leq k < m$. Then $g_{i-1}...g_1(U) \subset (U \setminus (W_1 \cup ... \cup W_{i-1})) \cup (g_{i-1}...g_1(U \cap W_1) \cup g_{i-1}...g_2(W_2 \cap U)... \cup g_{i-1}(U \cap W_{i-1}))$ and $W_i \cup g_{i-1}...g_1(U) \neq M$ for each $U \in \mathcal{U}$, since $U \neq M$ and $g(U) \neq M$, consequently, Condition (iii) of $(E3)$ is also satisfied.

For each clopen compact V in M the group $G_V := \{g \in G : supp(g) \subset V\}$ is topologically perfect by Theorem 3. Moreover, $G = \bigcup_V G_V$, where the union is by all clopen compact subsets V in M. The group G is supplied with the C^α compact-open topology. Therefore, G is topologically perfect. In view of the Epstein Theorem 2.2.1 and Corollary 2.2.2 [5] the commutator group $[G,G]$ is topologically simple, since each $[G_V, G_V]$ is topologically simple. On the other hand, $[G,G]$ is dense in G, hence G is topologically simple. If M is locally compact and noncompact, then $Diff_c^\alpha(M)$ is everywhere dense in $Diff^\alpha(M)$, hence the latter group is topologically simple relative to its C^α compact-open topology.

5. Remark. If M is a compact manifold, $t \geq 0$, $\alpha \in \{t, [t]\}$, then $Diff^\alpha(M)$ is not algebraically simple being the projective limit of groups due to Theorem 3.6.

6. Theorem. *Let M be a C^α compact manifold, $t \geq 0$, $\alpha \in \{t, [t]\}$. Then each continuous automorphism of $Diff^\alpha(M)$ belongs to the group of homeomorphisms $Hom(M,M) =: Hom(M)$.*

Proof. Let ψ be a continuous automorphism of $Diff^\alpha(M)$. In view of Theorem 3.6 above or Theorem 2.5 [24] for each $v \in \mathbf{N}$, $v \geq s$, there exists the quotient mapping $\pi_v^* : Diff^\alpha(M) \to \Sigma_{b_v}$, where Σ_m is the symmetric group of the set $\{1,...,m\}$, b_v is the cardinality of the finite set $M_v = \pi_v(M)$, π_v^* is induced by the quotient mapping $\pi_v : \mathbf{K} \to \mathbf{K}/B(\mathbf{K}, 0, |\pi|^v)$ with the help of polynomial expansions by Formulas 3.2(4,5). Consider M embedded into $\mathbf{K}^n$. The set of diffeomorphisms f such that $f - id$ is a piecewise affine on balls of the covering of M is contained in $Diff^\alpha(M)$, consequently, $\pi_v^*(Diff^\alpha(M)) = S_{b_v}$ is the epimorphism. This also follows from Formulas 3.2(4 – 7). The automorphism ψ induces the automorphism ψ_v of $\pi_v^*(Diff^\alpha(M))$ such that $\pi_v^*(\psi(g)) =: \psi_v(\pi_v^*(g))$, since then $\pi_v^*(\psi(gh)) = \pi_v^*(\psi(g)\psi(h)) = \pi_v^*(\psi(g))\pi_v^*(\psi(h)) = \psi_v(\pi_v^*(g))\psi_v(\pi_v^*(h))$ for each $g,h \in Diff^\alpha(M)$. Consider v sufficiently large such that $b_v > 6$. In view of the Hölder Theorem 5.3.1 [19] the automorphism ψ_v is internal: $\psi_v(a) = h_v a h_v^{-1}$ for each $a \in \pi_v^*(Diff^\alpha(M))$, where $h_v \in \pi_v^*(Diff^\alpha(M))$ is a marked element. This defines the inverse sequence $\{h_v, \pi_l^v, \Lambda_s\}$ such that $\pi_l^v(h_v) = h_l$ for each

$v \geq l \in \Lambda_s$, where $\pi_l^v \circ \pi_v^* = \pi_l^*$ and $\pi_l^v : G_v \to G_l$ are algebraic epimorphisms of discrete groups $G_v := \pi_v^*(Diff^{\alpha}(M))$ for each $v \geq l \in \Lambda_s$, π_l^v are written without star for simplicity of notation. Each $h_v : M_v \to M_v$ is the homeomorphism and each $\psi_v : G_v \to G_v$ is the homeomorphism. A limit of an inverse mapping system of homeomorphic mappings is a homeomorphism by Proposition 2.5.10 [11]. Therefore, $h = pr - \lim\{h_v, \pi_l^v, \Lambda_s\}$ is the element of $Diff_w(M)$ (see Theorems 3.6, 3.7 above and Theorem 2.5 [24]), moreover, $\psi(g) = \lim\{h_v \pi_v^*(g) h_v^{-1}, \pi_l^v, \Lambda_s\} = hgh^{-1}$ for each $g \in Diff^{\alpha}(M)$. On the other hand, $Diff_w(M)$ is algebraically isomorphic with $Hom(M)$, hence $h \in Hom(M)$.

7. Remark. Theorem 6 is not true for a locally compact noncompact manifold M for the group $Diff_c^{\alpha}(M)$ of compactly supported diffeomorphisms of $Diff^{\alpha}(M)$, since then $Diff_c^{\alpha}(M)$ has the external automorphisms $\phi_f(g) := fgf^{-1}$ for $f \in Diff^{\alpha}(M) \setminus Diff_c^{\alpha}(M)$, where $g \in Diff_c^{\alpha}(M)$. Moreover, $Diff_c^{\alpha}(M)$ is the proper normal subgroup in $Diff^{\alpha}(M)$, but $Diff_c^{\alpha}(M)$ for a locally compact noncompact manifold M is the proper non-closed subgroup everywhere dense in $Diff^{\alpha}(M)$.

7. Projective Decomposition of Wrap Groups

1. Let as in §2.1 $\bar{M}$ and N be two compact manifolds over a locally compact non-archimedean infinite field $\mathbf{K}$ with a multiplicative non trivial norm relative to which $\mathbf{K}$ is complete as the uniform space and $Diff_0^{\alpha}(\bar{M})$ be a subgroup in $Diff^{\alpha}(\bar{M})$ of all elements $\psi \in Diff^{\alpha}(\bar{M})$ such that $\psi(s_0) = s_0$, where s_0 is a marked point in $\bar{M}$, $\alpha \in \{\underline{t}, [t]\}$. Denote by $C_0^{\alpha}(M,N)$ a subspace in $C^{\alpha}(\bar{M},N)$ of all elements $f \in C^{\alpha}(\bar{M},N)$ such that $\lim_{|\zeta_1|+\dots+|\zeta_n| \to 0} \bar{\Phi}^v(f - w_0)(s_0; h_1, \dots, h_n; \zeta_1, \dots, \zeta_n) = 0$ for $\alpha = t$ or $\lim_{|\zeta_1|+\dots+|\zeta_n| \to 0} \Upsilon^v(f - w_0)(s_0^{[n]}) = 0$ for $\alpha = [t]$ for each $v \in \{0, 1, \dots, t\}$, where $M = \bar{M} \setminus s_0$ and $w_0(\bar{M}) = \{y_0\}$, $x^{[k+1]} = (x^{[k]}, v^{[k]}, \zeta_k)$ (see §2.6 [23] and Section 2 above). Geometric wrap monoids $\Omega_{\alpha}(M,N)$ and wrap groups $L_{\alpha}(M,N)$ for C^{α} classes of mappings were constructed in [23]. The same construction is for $char(\mathbf{K}) = p > 0$.

Theorem. *Let $\Omega_{\alpha}(M,N)$ be a commutative wrap monoid, then the quotient mappings π_k induce the corresponding inverse sequence $\{\Omega(M_k, N_k) : k \in \mathbf{N}\}$ such that $\Omega^w(M,N) := pr - \lim_k \Omega(M_k, N_k)$ is the commutative compact topological monoid, where $\pi_k^* : \Omega_{\alpha}(M,N) \to \Omega(M_k, N_k)$, $\pi_k^l : \Omega(M_l, N_l) \to \Omega(M_k, N_k)$ are surjective mappings for each $l \geq k$, $\Omega(M_k, N_k) = \{f_k : f_k \in N_k^{M_k}, f_k(s_{0,k}) =$*

$y_{0,k}\}/K_{\alpha,k}$, $K_{\alpha,k}$ is an equivalence relation induced by an equivalence relation K_α. Moreover, $\Omega^w(M,N)$ is a compactification of $\Omega_\alpha(M,N)$ relative to the projective weak topology τ_w.

Proof. In view of Corollary 3.3 $\pi_k(C_0^\alpha(M,N))$ is isomorphic with $\{f_k : f_k \in N_k^{M_k}, f_k(s_{0,k}) = y_{0,k}\}$, where the quotient mapping is denoted by π_k both for M and N, since it is induced by the same ring homomorphism $\pi_k : \mathbf{K} \to \mathbf{K}/B(\mathbf{K},0,1)$, $s_{0,k} := \pi_k(s_0)$ and $y_{0,k} := \pi_k(y_0)$, where $k \ge s = \max(s(M), s(N))$. Then $\pi_k^*(Diff_0^\alpha(M))$ is isomorphic with $Hom_0(M_k) := \{\psi_k : \psi_k \in Hom(M_k), \psi_k(s_{0,k}) = s_{0,k}\}$. All of this is also applicable with the corresponding changes to classes of smoothness C^α (or $\underline{C}(\alpha)$ in the notation of [23]), where $\alpha = t$ or $\alpha = [t]$ with substitution of Φ^k on Υ^k in the latter case. If f and g are two K_α-equivalent elements in $C_0^\alpha(M,N)$, that is, there are sequences f_n and g_n in $C_0^\alpha(M,N)$ converging to f and g respectively and also a sequence $\psi_n \in Diff_0^\alpha(M)$ such that $f_n(x) = g_n(\psi_n(x))$ for each $x \in M$, then $\pi_k^*(f_n) =: f_{n,k}$ and $g_{n,k} := \pi_k^*(g_n)$ converge to $\pi_k^*(f)$ and $\pi_k^*(g)$ respectively and also $\psi_{n,k} := \pi_k^*(\psi_n) \in Hom_0(M_k)$. From the equality $f_{n,k}(x(k)) = g_{n,k}(\psi_{n,k}(x(k)))$ for each $n \in \mathbf{N}$ and $x(k) \in M_k$ it follows, that the equivalence relation K_α induces the corresponding equivalence relation $K_{\alpha,k}$ in $\pi_k^*(C_0^\alpha(M,N))$ such that classes $< \pi_k^*(f) >_{K,\alpha,k}$ of $K_{\alpha,k}$-equivalent elements are closed. Each element $f_k \in \pi_k^*(C_0^\alpha(M,N))$ is characterized by the equality $f_k(s_{0,k}) = y_{0,k}$. This induces the quotient mapping $\pi_k^* : \Omega_\alpha(M,N) \to \Omega(M_k,N_k)$ and surjective mappings $\pi_k^l : \Omega(M_l,N_l) \to \Omega(M_k,N_k)$ for each $l \ge k$. Each $\Omega(M_k,N_k)$ is the finite discrete set, since each $N_k^{M_k}$ is the finite discrete set. This produces the inverse sequence of finite discrete spaces, hence the limit $\Omega^w(M,N) := pr-\lim\{\Omega(M_k,N_k), \pi_l^k, \Lambda_s\}$ of the inverse sequence is compact and totally disconnected. It remains to verify that $\Omega^w(M,N)$ is the commutative topological monoid with the unit element and the cancelation property.

From the equality $M = \bar{M} \setminus \{s_0\}$, it follows that $M_k = \bar{M}_k$, since for each $k \in \mathbf{N}$ there exists $x \in M$ such that $x + B(\mathbf{K}^m, 0, p^{-k}) \ni s_0$. Moreover, M_k and N_k are finite discrete spaces. Then $\pi_k(M \vee M) = M_k \vee M_k$, where $A \vee B := A \times \{b_0\} \cup \{a_0\} \times B \subset A \times B$ is the wedge product of pointed spaces (A, a_0) and (B, b_0), A and B are sets with marked points $a_0 \in A$ and $b_0 \in B$. The composition operation is defined on threads $\{< f_k >_{K,\alpha,k} : k \in \mathbf{N}\}$ of the inverse sequence in the following way. There was fixed a $C^{[\infty]}$-diffeomorphism $\chi : M \vee M \to M$ [23]. Let $x \in M$, then $\pi_k(x) \in M_k$ and $\chi^{-1}(U) \in M \vee M$, where $U := \pi_k^{-1}(x + B(\mathbf{K}, 0, p^{-k})) \cap M$. On the other hand $\chi^{-1}(U)$ is a disjoint union of balls of radius p^{-2k} in $B(\mathbf{K}^{2m}, 0, 1)$, hence there is defined a surjective mapping χ_k :

$M_{2k}\vee M_{2k}\to M_k$ induced by χ, π_k and π_{2k} such that $\chi_k(\chi^{-1}(U))=\pi_k(x)$. If f and $g\in C^\alpha(M,N)$, then $f\vee g\in C^\alpha((M\vee M),N)$ and $\chi(f\vee g)\in C^\alpha(M,N)$ as in §2.6 [23]. Hence $\chi_k(f_{2k}\vee g_{2k})\in C^\alpha(M_k,N_k)$ and inevitably $\chi_k(<f_{2k}\vee g_{2k}>_{K,\alpha,2k})=\chi_k(<f_{2k}>_{K,\alpha,2k}\vee<g_{2k}>_{K,\alpha,2k})\in\Omega(M_k,N_k)$.

There exists a one to one correspondence between elements $f\in C_w(\bar{M},N)$ and $\{f_k:k\}\in\{N_k^{M_k}:k\in\Lambda_s\}$. Therefore, $pr-\lim_k\Omega(M_k,N_k)$ algebraically this is the commutative monoid with the cancelation property. Let U be a neighborhood of e in $\Omega^w(M,N)$, then there exists $U_k=\pi_k^{-1}(V_k)$ such that V_k is open in $\Omega(M_k,N_k)$, $e\in U_k$ and $U_k\subset U$. On the other hand there exists $U_{2k}=\pi_{2k}^{-1}(V_{2k})$ such that V_{2k} is open in $\Omega(M_{2k},N_{2k})$, $e\in U_{2k}$ and $U_{2k}+U_{2k}\subset U_k$. Therefore, $(f+U_{2k})+(g+U_{2k})\subset f+g+U_k\subset f+g+U$ for each $f,g\in\Omega^w(M,N)$, consequently, the composition in $\Omega^w(M,N)$ is continuous. Since $C_0^\alpha(M,N)$ is dense in $C_{0,w}(\bar{M},N)$, then $\Omega_\alpha(M,N)$ is dense in $\Omega^w(M,N)$ relative to the projective weak topology τ_w.

2. Corollary. *The wrap group $L_\alpha(M,N)$ has a non-archimedean compactification $L^w(M,N)$ relative to the projective weak topology τ_w.*

Proof. Using the Grothendieck construction we get a compactification $L^w(M,N)=\bar{F}/\bar{B}$ of a wrap group $L_\alpha(M,N)$, where $\bar{F}$ is a closure in $(\Omega^w(M,N))^{\mathbf{Z}}$ of a free commutative group F generated by $\Omega^w(M,N)$ and $\bar{B}$ is a closure of a subgroup B generated by all elements $[a+b]-[a]-[b]$, since the product of compact spaces is compact by the Tychonoff theorem.

3. Let now $s_0=0$ and $y_0=0$ be two marked points in the compact manifolds $\bar{M}$ and N embedded into $\mathbf{K}^{\mathsf{m}}$ and $\mathbf{K}^{\mathsf{n}}$ respectively. There is defined the following $C^{[\infty]}$-diffeomorphism $inv:(\mathbf{K}^{\mathsf{m}})'\to(\mathbf{K}^{\mathsf{m}})'$ for $(\mathbf{K}^{\mathsf{m}})':=\mathbf{K}^{\mathsf{m}}\setminus\{x:$ there exists j with $x_j=0\}$ such that $inv(x_1,\dots,x_{\mathsf{m}})=(x_1^{-1},\dots,x_{\mathsf{m}}^{-1})$. Let $M'=M\cap(\mathbf{K}^{\mathsf{m}})'$, then $inv(M')$ is locally compact and unbounded in $\mathbf{K}^{\mathsf{m}}$, consequently, $\pi_k(inv(M'))=(inv(M'))_k$ is a discrete infinite subset in $\mathbf{K}_{\mathbf{k}}^{\mathsf{m}}$ for each $k\in\mathbf{N}$. Analogously $\pi_k(inv(M'\vee M'))=(inv(M'\vee M'))_k\subset\mathbf{K}_{\mathbf{k}}^{2\mathsf{m}}$. There exists a $C^{[\infty]}$-diffeomorphism $\chi:M\vee M\to M$ such that $inv\circ\chi\circ inv$ is the $C^{[\infty]}$-diffeomorphism of $inv(M'\vee M')$ with $inv(M')$ and it induces bijective mappings χ_k of $inv((inv(M'\vee M'))_k)$ with $inv((inv(M'))_k)$ for each $k\in\mathbf{N}$ such that $\hat{\pi}_k^l\circ\chi_l=\chi_k$ for each $l\ge k$, where $\hat{\pi}_k^l:=inv\circ\pi_k^l\circ inv$. This produces inverse sequences of discrete spaces $inv((inv(M'))_k)=:\hat{M}_k$, $inv((inv(M'\vee M'))_k)=\hat{M}_k\vee\hat{M}_k$ and their bijections χ_k such that $pr-\lim_k\hat{M}_k$ is homeomorphic with M' and $pr-\lim_k\chi_k$ is equal to χ up to the homeomorphism, since $pr-\lim_k\mathbf{K}_{\mathbf{k}}^{\mathsf{m}}=\mathbf{K}^{\mathsf{m}}$ (see also about admissible modifications and polyhedral expansions in [27, 28]). If

$\psi \in Diff_0^\alpha(\bar{M})$, then $\hat{\psi} \in Diff^\alpha(\hat{M})$. Let $J_{f,k} := \{h_k : h_k = f_k \circ \psi_k, \psi_k \in Hom(\hat{M}_k), \psi_k(s_{0,k}) = s_{0,k}\}$ for $f_k \in N_k^{\hat{M}_k}$ with $\lim_{x\to 0} f_k(x) = 0$, then $J_{f,k}$ is closed and $\hat{\pi}_k^*(< f >_{K,\alpha}) \subset J_{f,k}$. Therefore, g_k and f_k are $\hat{K}_{\alpha,k}$-equivalent if and only if there exists $\psi_k \in Hom(\hat{M}_k)$ such that $\psi_k(s_{0,k}) = s_{0,k}$ and $g_k(x) = f_k(\psi_k(x))$ for each $x \in \hat{M}_k$. Let $\Omega(\hat{M}_k, N_k) := \hat{\pi}_k^*(\Omega_\alpha(M,N))$.

Theorem. *The set of $\Omega(\hat{M}_k, N_k)$ forms an inverse sequence $S = \{\Omega(\hat{M}_k, N_k); \hat{\pi}_k^l; k \in \Lambda_s\}$ such that $pr-\lim S =: \Omega^{i,w}(M,N)$ is an associative topological wrap monoid with the cancelation property and the unit element e. There exists an embedding of $\Omega_\alpha(M,N)$ into $\Omega^{i,w}(M,N)$ such that $\Omega_\alpha(M,N)$ is dense in $\Omega^{i,w}(M,N)$ relative to the projective weak topology $\tau_{i,w}$.*

Proof. Let U'_i be an analytic disjoint atlas of $inv(M')$, $f \in C^\alpha(inv(M'), \mathbf{K})$, $\psi \in Diff^\alpha(inv(M'))$, then each restriction $f|_{U'_i}$ has the form $f|_{U'_i}(x) = \sum_m f_{i,m} \bar{Q}_{i,m}(x)$ for each $x \in U'_i$, where $\bar{Q}_{i,m}$ are basic Amice polynomials for U'_i, $f_{i,m} \in \mathbf{K}$. Therefore f is a combination $f = \nabla_i f|_{U'_i}$, hence $\hat{\pi}_k^*(f \circ \psi(x)) = \sum_m [\hat{\pi}_k^*(f_{i,m}) \nabla_{(i,\psi_k(x(k))\in\hat{\pi}_k(U'_k)} \bar{Q}_{i,m,k}(\psi_k(x(k)))]$ and inevitably $\hat{\pi}_k^*((f \circ \psi)(x)) = f_k \circ \psi_k(x(k))$, where $Q_{i,m,k} := \hat{\pi}_k^*(Q_{i,m})$, $x \in inv(M')$ and $x(k) = \hat{\pi}_k(x)$.

As in §2.6.2 [23] we choose an infinite atlas $At'(M) := \{(U'_j, \phi'_j) : j \in \mathbf{N}\}$ such that $\phi'_j : U'_j \to B(X, y'_j, r'_j)$ are homeomorphisms, $\lim_{k\to\infty} r'_{j(k)} = 0$, $\lim_{k\to\infty} y'_{j(k)} = 0$ for an infinite sequence $\{j(k) \in \mathbf{N} : k \in \mathbf{N}\}$ such that $cl_{\bar{M}}[\bigcup_{k=1}^\infty U'_{j(k)}]$ is a clopen neighborhood of zero in $\bar{M}$, where $cl_{\bar{M}}A$ denotes the closure of a subset A in $\bar{M}$. We take $|y'_{j(k)}| > r'_{j(k)}$ for each k, hence $inv(B(X, y'_j, r'_j) \cap X') = B(X, y'^{-1}_j, r'^{-1}_j) \cap X'$ and $\bigcup_k inv(U'_{j(k)} \cap X')$ is open in X', where $X = \mathbf{K}^m$. For an atlas $At'(M \vee M) := \{(W_l, \alpha_l) : l \in \mathbf{N}\}$ with homeomorphisms $\alpha_l : W_l \to B(X, z_l, a_l)$, $\lim_{k\to\infty} a_{l(k)} = 0$, $\lim_{k\to\infty} z_{l(k)} = 0$ for an infinite sequence $\{l(k) \in \mathbf{N} : k \in \mathbf{N}\}$ such that $cl_{\bar{M}\vee\bar{M}}[\bigcup_{k=1}^\infty W_{l(k)}]$ is a clopen neighborhood of 0×0 in $\bar{M} \vee \bar{M}$ we also choose $|z_l| > a_l$ for each l, where $card(\mathbf{N} \setminus \{l(k) : k \in \mathbf{N}\}) = card(\mathbf{N} \setminus \{j(k) : k \in \mathbf{N}\})$.

Then we take $\chi(W_{l(k)}) = U'_{j(k)}$ for each $k \in \mathbf{N}$ and $\chi(W_l) = U'_{\kappa(l)}$ for each $l \in (\mathbf{N} \setminus \{l(k) : k \in \mathbf{N}\})$, where $\kappa : (\mathbf{N} \setminus \{l(k) : k \in \mathbf{N}\}) \to (\mathbf{N} \setminus \{j(k) : k \in \mathbf{N}\})$ is a bijective mapping such that $p^{-1} \le r'_{j(k)}/a_{l(k)} \le p$ for each k and $p^{-1} \le r'_{\kappa(l)}/a_l \le p$ for each $l \in (\mathbf{N} \setminus \{l(k) : k \in \mathbf{N}\})$. We can choose the locally affine mapping χ on $M = \bar{M} \setminus \{s_0\}$ such that $\Phi^n \chi = 0$ or $\Upsilon^n \chi = 0$ for each $n \ge 2$ and $B(X', y'^{-1}_l, r'^{-1}_l)$ are diffeomorphic with $inv(U'_l \cap X')$ and $B(X' \vee X', z_l^{-1}, a_l^{-1})$ are diffeomorphic with $inv(W_l \cap (X' \vee X'))$.

This induces the diffeomorphisms $\hat{\chi} := inv \circ \chi \circ inv : \hat{M} \vee \hat{M} \to \underline{\hat{M}}$ and $\hat{\chi}^* : C_0^\alpha((\hat{M} \vee \hat{M}, \infty \times \infty), (N, y_0)) \to C_0^\alpha((\hat{M}, \infty), (N, y_0))$, since each $\Phi^n(f \vee$

$g)(\hat{\chi}^{-1})$ or $\Upsilon^n(f\vee g)(\hat{\chi}^{-1})$ has an expression through $\bar{\Phi}^l(f\vee g)$ and $\bar{\Phi}^j(\hat{\chi}^{-1})$ or $\Upsilon^l(f\vee g)$ and $\Upsilon^j(\hat{\chi}^{-1})$ respectively with $l,j\le q$ and q subordinated to α, where $\hat{M}:=inv(M')$ and conditions defining the subspace $C_0^\alpha((\hat{M},\infty),(N,y_0))$ differ from that of $C_0^\alpha((M,s_0),(N,y_0))$ by substitution of $\lim_{x\to s_0}$ on $\lim_{|x|\to\infty}$. Then $\lim_{|x|\to\infty}|\hat{\chi}(x)|=\infty$, consequently, there exists $k_0\in\mathbf{N}$ such that $\hat{\chi}_k:\hat{M}_k\vee\hat{M}_k\to\hat{M}_k$ are bijections for each $k\ge k_0$, where $\hat{\chi}_k:=\hat{\pi}_k\circ\hat{\chi}$. If $\psi\in Diff^\alpha(\bar{M})$ and $\psi(0)=0$, then $\lim_{|x|\to\infty}\hat{\psi}(x)=\infty$ and $\lim_{|x|\to\infty}\hat{\psi}^{-1}(x)=\infty$. Then considering $\hat{\psi}_k$ we get an equivalence relation $K_{\alpha,k}$ in $\{f_k: f_k\in N_k^{\hat{M}_k},\lim_{|x|\to\infty}f_k(x)=0\}$ induced by K_α, where $\hat{M}_k$ is supplied with the quotient norm induced from the space X, since $X'\subset X$, $x\in\hat{M}_k$.

Let J_k denotes the quotient mapping corresponding to $K_{\alpha,k}$. Therefore analogously to §2.6 [23] we get, that $\Omega(\hat{M}_k,N_k)$ are commutative monoids with the cancelation property and the unit elements e_k, since $\Omega(\hat{M}_k,N_k)=\{f_k: f_k\in C^0(\hat{M}_k,N_k),\lim_{|x|\to\infty}f_k(x)=0\}/\hat{K}_{\alpha,k}$ and mappings $\hat{\pi}_k^l:(\mathbf{K}^{\mathsf{m}})'_l\to(\mathbf{K}^{\mathsf{m}})'_k$ and mappings $\pi_k^l:\mathbf{K}^{\mathsf{n}}{}_l\to\mathbf{K}^{\mathsf{n}}{}_k$ induce mappings $\hat{\pi}_k^l:\Omega(\hat{M}_l,N_l)\to\Omega(\hat{M}_k,N_k)$ for each $l\ge k$. Let the topology in $\{f_k: f_k\in C^0(\hat{M}_k,N_k),\lim_{|x|\to\infty}f_k(x)=0\}$ be induced from the Tychonoff product topology in $N_k^{\hat{M}_k}$ and $\Omega(\hat{M}_k,N_k)$ be in the quotient topology.

The space $N_k^{\hat{M}_k}$ is metrizable by the Baire metric $\rho(x,y):=|\pi|^{-j}$, where $j=\min\{i: x_i\ne y_i, x_1=y_1,\dots,x_{i-1}=y_{i-1}\}$, $x=(x_l: x_l\in N_k, l\in\mathbf{N})$, $\hat{M}_k$ is enumerated as $\mathbf{N}$, $\pi\in\mathbf{K}$, $0<|\pi|<1$ is the generator of the valuation group $\Gamma_{\mathbf{K}}$. Therefore, $\Omega(\hat{M}_k,N_k)$ is metrizable and the mapping $(f_k,g_k)\to f_k\vee g_k$ is continuous, hence the mapping $(J_k(f_k),J_k(g_k))\to J_k(f_k)\circ J_k(g_k)$ is also continuous. Then $J_k(w_{0,k})$ is the unit element, where $w_{0,k}(\hat{M}_k)=0$. Hence $\Omega^{i,w}(M,N)$ is the commutative monoid with the cancelation property and the unit element. Certainly $\prod_k\Omega(\hat{M}_k,N_k)$ is the topological monoid and $pr-\lim S$ is a closed in it topological totally disconnected monoid. For each $f\in C_0^\alpha(M,N)$ there exists an inverse sequence $\{f_k: f_k=\hat{\pi}_k^*(f),k\in\Lambda_s\}$ such that $f(x)=pr-\lim_k f_k(x(k))$ for each $x\in M'$, but M' is dense in M. Therefore there exists an embedding $\Omega^\alpha(M,N)\hookrightarrow\Omega^{i,w}(M,N)$, hence $\Omega^\alpha(M,N)$ is dense in $\Omega^{i,w}(M,N)$ relative to the projective weak topology $\tau_{i,w}$, since $C^\alpha(M,N)$ is dense in $C_w(M,N)$ relative to the τ_w topology.

4. Corollary. *The inverse sequence of wrap monoids induces the inverse sequence of wrap groups $S_L:=\{L(\hat{M}_k,N_k);\hat{\pi}_k^l;\Lambda_s\}$. Its projective limit $L^{i,w}(M,N):=pr-\lim S_L$ is a commutative topological totally disconnected* *group and $L_\alpha(M,N)$ has an embedding in it as a dense subgroup.*

Proof. Due to the Grothendieck construction the inversion operation $f_k \mapsto f_k^{-1}$ is continuous in $L(\hat{M}_k,N_k)$ and homomorphisms $\hat{\pi}_k^l$ and $\hat{\pi}_k$ have continuous extensions from wrap submonoids onto wrap groups $L(\hat{M}_k,N_k)$. Each monoid $\Omega(\hat{M}_k,N_k)$ is totally disconnected, since $N_k^{\hat{M}_k}$ is totally disconnected and $\Omega(\hat{M}_k,N_k)$ is supplied with the quotient ultrametric, hence the free Abelian group F_k generated by $\Omega(\hat{M}_k,N_k)$ is also totally disconnected and ultramertizable, consequently, $L(\hat{M}_k,N_k)$ is ultrametrizable. Evidently their inverse limit is also ultrametrizable and the equivalent ultrametric can be chosen with values in $\tilde{\Gamma}_{\mathbf{K}} := \{|z| : z \in \mathbf{K}\}$, where $\tilde{\Gamma}_{\mathbf{K}} \cap (0,\infty)$ is discrete in $(0,\infty) := \{x : 0 < x < \infty, x \in \mathbf{R}\}$. Then the projective limit (that is, weak) topology of $L^{i,w}(M,N)$ is induced by the projective weak topology of $C_w(M,\mathbf{K})$.

5. Theorem. *For each prime number p the wrap group $L_\alpha(M,N)$ in its weak topology inherited from $L^{i,w}(M,N)$ has the non-archimedean compactification isomorphic with $\mathbf{Z_p}^{\aleph_0}$, moreover, $L^{i,w}(M,N)$ has the compactification $(\mathrm{v}\mathbf{Z})^{\aleph_0}$, where $\mathrm{v}\mathbf{Z}$ is the one-point Alexandroff compactification of $\mathbf{Z}$.*

Proof. The projective ring homomorphism $\pi_k : \mathbf{K} \to \mathbf{K_k}$ induces
$\hat{\pi}_k^*(\bar{\Phi}^m(f(x;h_1,...,h_m;\zeta_1,...,\zeta_m)) = \bar{\Phi}^m f_k(x(k);h_1(k),...,h_m(k);\zeta_1(k),...,\zeta_m(k))$
and $\hat{\pi}_k^*(\Upsilon^m(f(x^{[m]})) = \Upsilon^{[m]} f_k(x(k)^{[m]})$,
where $m \in \mathbf{N}$, $\bar{\Phi}^m f_k$ and $\Upsilon^m f_k$ are defined for the field of fractions generated by $\mathbf{K_k}$, since $\mathbf{K}$ is the commutative field (see also [4] and §§2.1-2.6 [23]). Then the condition

$$\lim_{|x|\to\infty} \bar{\Phi}^m f(x;h_1,...,h_m;\zeta_1,...,\zeta_m) = 0 \text{ or}$$

$$\lim_{|x|\to\infty} \Upsilon^m f(x^{[m]}) = 0$$

implies the condition

$$\lim_{|x(k)|\to\infty} \bar{\Phi}^m f_k(x(k);h_1(k),...,h_m(k);\zeta_1(k),...,\zeta_m(k)) = 0 \text{ or}$$

$$\lim_{|x(k)|\to\infty} \Upsilon^m f_k(x(k)^{[m]}) = 0$$

respectively, where $x^{[1]} = (x,v^{[0]},\zeta_1)$, $x^{[m+1]} := (x^{[m]},v^{[m]},\zeta_{m+1})$. Therefore, $supp(f_k) := \hat{M}_k^f := \{x(k) : f_k(x(k)) \neq 0\}$ is a finite subset of the discrete space $\hat{M}_k$ for each $k \in \mathbf{N}$. Then evidently, $\hat{\pi}_k^*(< g >_{K,\alpha})$ is a closed subset in $N_k^{\hat{M}_k}$ for each $g \in C_0^{\alpha}((\hat{M},\infty),(N,0))$, since for each limit point f_k of $\hat{\pi}_k^*(< g >_{K,\alpha})$ its support is the finite subset in $\hat{M}_k$. Let k_0 be such that $N_{k_0} \ne \{0\}$, then this is also true for each $k \ge k_0$. If $f_k \notin \hat{\pi}_k^*(< w_0 >_{K,\alpha})$ and $k \ge k_0$, then $f_k^{\vee n} \notin \hat{\pi}_k^*(< w_0 >_{K,\alpha}$) for each $n \in \mathbf{N}$, where $f_k^{\vee n} := f_k \vee ... \vee f_k$ denotes the n-times wedge product, since $\|f^{\vee n}\|_{C^\alpha} \ge \|f\|_{C^\alpha} > 0$ and $\|f_k^{\vee n}\|_{C(\mathbf{K_k^m},\mathbf{K_k^n})} \ge \|f\|_{C(\mathbf{K_k^m},\mathbf{K_k^n})} > 0$, where $C(\mathbf{K_k^m},\mathbf{K_k^n}) = \pi_k^*(C_b^\alpha(\mathbf{K^m},\mathbf{K^n}))$ is the quotient module over the ring $\mathbf{K_k}$. Each $\hat{\pi}_k^*(< f >_{K,\alpha})$ can be presented as the following composition $z_1 b_1 + ... + z_l b_l$ in the additive group $L(\hat{M}_k,N_k)$, where each b_i corresponds to $\hat{\pi}_k^*(< g_i >_{K,\alpha})$ and the embedding of $\Omega(\hat{M}_k,N_k)$ into $L(\hat{M}_k,N_k)$, $z_i \in \{-1,0,1\}$, $l = card(\hat{M}_k^f)$, $\hat{M}_k^{g_i}$ are singletons for each $i = 1,...,l$.

Each $\hat{M}_k$ is the finite discrete set as well as N_k. For each $x \ne y \in \hat{M}_k$ there exists $\psi \in Hom_0(M_k)$ such that $\psi(x) = y$, where 0 corresponds to s_0 for convenience of the notation. Using the group $Hom_0(N_k)$ we get that $L(\hat{M}_k,N_k)$ is isomorphic with $\mathbf{Z}^{n_k}$, where $n_k = card(N_k) > 1$. For each prime number $p > 1$ there exists the p-adic completion of $\mathbf{Z}$ which is $\mathbf{Z_p}$. In view of Corollary 4 $L_\alpha(M,N)$ has the non-archimedean completion isomorphic with $\mathbf{Z}_p^{\aleph_0}$, since $\mathbf{Z}$ is dense in $\mathbf{Z_p}$ and $pr-\lim_k \mathbf{Z}^{n_k} = \mathbf{Z}^{\aleph_0}$.

On the other hand, we can take the multiplicative subgroup $\{\theta^l : l \in \mathbf{Z}\}$ of the locally compact field $\mathbf{F_{p^u}}(\theta)$ which gives the embedding ϕ of $\mathbf{Z}$ into $\mathbf{F_{p^u}}(\theta)$, where $\theta^0 = 1$. The completion of $\phi(\mathbf{Z})$ in $\mathbf{F_{p^u}}(\theta)$ is $\phi(\mathbf{Z}) \cup \{0\}$ which is the one-point Alexandroff compactification $\nu\mathbf{Z}$ of $\mathbf{Z}$. This gives the non-archimedean completion $(\nu\mathbf{Z})^{\aleph_0}$ of $L_\alpha(M,N)$. Moreover, $\mathbf{Z}_p^{\aleph_0}$ and $(\nu\mathbf{Z})^{\aleph_0}$ are compact as products of compact spaces and $L^{i,w}(M,N)$ has the aforementioned embeddings into them.

6. Note. Using quotient mappings $\eta_{p,s} : \mathbf{Z} \to \mathbf{Z}/p^s\mathbf{Z}$ we get that $L_\alpha(M,N)^{\aleph_0}$ has the compactification equal to $\{\prod_{p \in \mathsf{P}} \mathbf{Z_p}^{\aleph_0}\} \times (\nu\mathbf{Z})^{\aleph_0}$ relative to the product Tychonoff topology, where P denotes the set of all prime numbers $p > 1$, $s \in \mathbf{N}$. These compactifications produce characters of $L_{i,w}(M,N)$, since each compact Abelian group has only one-dimensional irreducible unitary representations [16]. On the other hand, there are irreducible continuous representations of compact groups in non-archimedean Banach spaces [39]. Among them there are infinite-dimensional [10, 37]. Moreover, in their initial C^α topologies diffeomorphism and wrap groups also have infinite-dimensional irreducible unitary representations [22, 23]. At the same time topologies of $L_\alpha(M,N)$ and $L^w(M,N)$ or $L^{i,w}(M,N)$ are incomparable, since the topologies of $C^\alpha(M,N)$ and $C_w(M,N)$ are incomparable (see Theorem 3.7 above).

Projective limits of groups obtained above have the non-archimedean origin related with non-archimedean families of semi-norms on spaces of continuous or more narrow classes of functions between manifolds over ultra-normed fields. Generally, if a topological space X has a projective limit decomposition $X = pr - \lim\{X_\alpha, \pi^\alpha_\beta, \Lambda\}$, then if $f_\beta : X_\beta \to Y$ is a continuous function into a topological space Y, then $f := f_\beta \circ \pi_\beta : X \to Y$ is a continuous function, where the mapping $\pi^\alpha_\beta : X_\alpha \to X_\beta$ is continuous for each $\alpha \geq \beta \in \Lambda$, Λ is a directed set, $\pi^\alpha_\beta \circ \pi_\alpha = \pi_\beta$, $\pi_\alpha : X \to X_\alpha$ is continuous and epimorphic. Therefore, $f_\alpha = \tilde{\pi}^\beta_\alpha(f_\beta) := f_\beta \circ \pi^\alpha_\beta$ for each $\alpha \geq \beta$ generate the inductive limit $ind - \lim\{C(X_\beta, Y); \tilde{\pi}^\beta_\alpha; \Lambda\}$, where $C(X,Y)$ denotes the family of all continuous mappings from X into Y. On the other hand, if $Y = pr - \lim\{Y_\gamma, p^\gamma_\delta, \Psi\}$, then one gets $pr - \lim\{C(X, Y_\gamma); p^\gamma_\delta, \Psi\}$. Then these two constructions can be combined with repeated application of projective and inductive limits, which may be dependent on the order of taking limits. If $card(\Psi) \geq \aleph_0$, then a suitable box topology in $\prod_{\gamma\in\Psi} C(X, Y_\gamma)$ is strictly stronger than a weak topology in it and in its projective limit subspace (see also [32]).

7. Theorem. *If $\bar{M}$ and N are compact manifolds, $\alpha \in \{\infty, [\infty]\}$, then $L_\alpha(M,N)$ is the C^α Lie group.*

Proof. The uniform space $C^\alpha((M,s_0),(N,y_0))$ has the structure of the C^α manifold, since M and N are C^α manifolds, where $\alpha \in \{\infty, [\infty]\}$. Therefore, $\Omega_\alpha(M,N)$ and $L_\alpha(M,N)$ are C^α manifolds. The wedge product $(f,g) \mapsto f \vee g$ in $C^\alpha((M,s_0),(N,y_0))$ is the C^α mapping, since $M = \bar{M} \setminus \{s_0\}$. Using the quotient mapping by closures of equivalence relation caused by the action of $Diff^\alpha_0(M)$ becomes the C^α manifold and C^α monoid with the C^α smooth composition. Using the construction of $L_\alpha(M,N)$ we get, that $L_\alpha(M,N)$ is the C^α Lie group (see also for more details [23]).

8. Remark. Theorem 7 can be generalized in the C^α_b class for noncompact C^α_b manifolds M and N.

Appendix

1. Lemma. *Let either $f, g \in C^{[n]}(U,Y)$, where U is an open subset in X, Y is an algebra over* **K**, *or $f \in C^{[n]}(U, \mathbf{K})$ and $g \in C^{[n]}(U,Y)$, where Y is a topological vector space over* **K**, *then*

(1) $(fg)^{[n]}(x^{[n]}) = (\Upsilon\otimes\hat{P}+\hat{\pi}\otimes\Upsilon)^n.(f\otimes g)(x^{[n]})$
and $(fg)^{[n]} \in C^0(U^{[n]},Y)$, *where* $(\hat{\pi}^k g)(x^{[k]}) := g\circ\pi_1^0\circ\pi_2^1\circ\ldots\circ\pi_k^{k-1}(x^{[k]})$, $\hat{P}^n g := P_nP_{n-1}\ldots P_1 g$, $\pi_k^{k-1}(x^{[k]}) := x^{[k-1]}$, $(A\otimes B).(f\otimes g) := (Af)(Bg)$ *for* $A,B\in L(C^n(U,Y),C^m(U,Y))$, $m\le n$, $(A_1\otimes B_1)\ldots(A_k\otimes B_k).(f\otimes g) := (A_1\ldots A_k\otimes B_1\ldots B_k).(f\otimes g) := (A_1\ldots A_kf)(B_1\ldots B_kg)$ *for corresponding operators,* $\Upsilon^n f := f^{[n]}$, $(P_kg)(x^{[k]}) := g(x^{[k-1]}+v^{[k-1]}t_k)$,
$\hat{P}^k\hat{\pi}^{a_1}\Upsilon^{b_1}\ldots\hat{\pi}^{a_l}\Upsilon^{b_l}g = P_{k+s}\ldots P_{s+1}\hat{\pi}^{a_1}\Upsilon^{b_1}\ldots\hat{\pi}^{a_l}\Upsilon^{b_l}g$ *with* $s=b_1+\ldots+b_l-a_1-\ldots-a_l\ge 0$, $a_1,\ldots,a_l,b_1,\ldots,b_l\in\{0,1,2,3,\ldots\}$.

Proof. Let at first $n=1$, then

(2) $(fg)^{[1]}(x^{[1]}) = [(fg)(x+vt)-(fg)(x)]/t = [(f(x+vt)-f(x))g(x+vt)+f(x)(g(x+vt)-g(x))]/t = (\Upsilon^1 f)(x^{[1]})(P_1g)(x^{[1]})+(\hat{\pi}_1^0 f)(x^{[1]})\Upsilon^1 g(x^{[1]})$,
since $\hat{\pi}_1^0(x^{[1]}) = x$ and P_1 is the composition of the projection $\hat{\pi}_1^0$ and the shift operator on vt. Let now $n=2$, then applying Formula (2) we get:

(3) $(fg)^{[2]}(x^{[2]}) = ((fg)^{[1]}(x^{[1]}))^{[1]}(x^{[2]}) = (\Upsilon^1(f^{[1]}(x^{[1]})(x^{[2]}))g(x+(v^{[0]}+v_2^{[1]}t_2)(t_1+v_3^{[1]}t_2)+v_1^{[1]}t_2) + f^{[1]}(x^{[1]})g^{[1]}(x+v^{[0]}t_1, v_1^{[1]}+v_2^{[1]}(t_1+v_3^{[1]}t_2),t_2) + f^{[1]}(x,v_1^{[1]},t_2)g^{[1]}(x^{[1]}+v_1^{[1]}t_2)+f(x)g^{[2]}(x^{[2]})$,
where $v^{[k]} = (v_1^{[k]},v_2^{[k]},v_3^{[k]})$ for each $k\ge 1$ and $v^{[0]} = v_1^{[0]}$ such that $x^{[k]}+v^{[k]}t_{k+1} = (x^{[k]}+v_1^{[k]}t_{k+1}, v^{[k-1]}+v_2^{[k]}t_{k+1}, t_k+v_3^{[k]}t_{k+1})$ for each $1\le k\in\mathbf{Z}$. For $n=3$ we get

(4) $(fg)^{[3]}(x^{[3]}) = [(\Upsilon^3 f)(\hat{P}^3 g) + (\hat{\pi}^1\Upsilon^2 f)(\Upsilon^1\hat{P}^2 g) + (\Upsilon^1(\hat{\pi}^1\Upsilon^1 f))(\hat{P}^1\Upsilon^1\hat{P}^1 g)$
$+(\hat{\pi}^2\Upsilon^1 f)(\Upsilon^2\hat{P}^1 g)+(\Upsilon^2\hat{\pi}^1 f)(\hat{P}^2\Upsilon^1 g)+(\hat{\pi}^1\Upsilon^1\hat{\pi}^1 f)(\Upsilon^1\hat{P}^1\Upsilon^1 g)$
$+(\Upsilon^1(\hat{\pi}^2 f))(\hat{P}^1\Upsilon^2 g)+(\hat{\pi}^3 f)(\Upsilon^3 g)](x^{[3]})$,
since by our definition $\hat{P}^k\hat{\pi}^{a_1}\Upsilon^{b_1}\ldots\hat{\pi}^{a_l}\Upsilon^{b_l}g = P_{k+s}\ldots P_{s+1}\hat{\pi}^{a_1}\Upsilon^{b_1}\ldots\hat{\pi}^{a_l}\Upsilon^{b_l}g$ with $s=b_1+\ldots+b_l-a_1-\ldots-a_l\ge 0$, $a_1,\ldots,a_l,b_1,\ldots,b_l\in\{0,1,2,3,\ldots\}$.

Therefore, Formula (1) for $n=1$ and $n=2$ and $n=3$ is demonstrated by Formulas $(2-4)$. If $f,g\in C^0(U^{[k]},Y)$, $a,b\in\mathbf{K}$, then $(P_k(af+bg))(x^{[k]}) := (af+bg)(x^{[k-1]}+v^{[k-1]}t_k) = af(x^{[k-1]}+v^{[k-1]}t_k)+bg(x^{[k-1]}+v^{[k-1]}t_k)$, moreover, $\hat{\pi}^k(af+bg)(x^{[k]}) = (af+bg)\circ\pi_1^0\circ\pi_2^1\circ\ldots\circ\pi_k^{k-1}(x^{[k]}) = (af+bg)(x) = af(x)+bg(x) = a\hat{\pi}^k f(x^{[k]})+b\hat{\pi}^k g(x^{[k]})$ for each $x^{[k]}\in U^{[k]}$, hence $\hat{\pi}^k$ and P_k and $\hat{P}^k$ are $\mathbf{K}$-linear operators for each $k\in\mathbf{N}$. Suppose that Formula (1) is proved for $n=1,\ldots,m$, then for $n=m+1$ it follows by application of Formula (2) to both sides of Formula (1) for $n=m$:

$(fg)^{m+1}(x^{[m+1]}) = ((fg)^{[m]}(x^{[m]}))^{[1]}(x^{[m+1]}) = ((\Upsilon\otimes\hat{P}+\hat{\pi}\otimes\Upsilon)^m.(f\otimes g)(x^{[m]}))^{[1]}(x^{[m+1]}) = (\Upsilon\otimes\hat{P}+\hat{\pi}\otimes\Upsilon)^{m+1}.(f\otimes g)(x^{[m+1]})$,
since $x^{[m+1]} = (x^{[m]})^{[1]}$ and more generally $x^{[m+k]} = (x^{[m]})^{[k]}$ for each nonnegative integers m and k such that $\pi_k^{k-1}(x^{[m+k]}) = x^{[m+k-1]}$ for $k\ge 1$; Υ^k, $\hat{P}^k$ and

$\hat{\pi}$ are **K**-linear operators on corresponding spaces of functions (see above and Lemma 2.3) and

$(\Upsilon\otimes\hat{P}+\hat{\pi}\otimes\Upsilon)^{m+1}.(f\otimes g)(x^{[m+1]})=$
$\sum_{a_1+...+a_{m+1}+b_1+...+b_{m+1}=m+1}(\Upsilon^{a_1}\otimes\hat{P}^{a_1})$
$(\hat{\pi}^{b_1}\otimes\Upsilon^{b_1})...(\Upsilon^{a_{m+1}}\otimes\hat{P}^{a_{m+1}})(\hat{\pi}^{b_{m+1}}\otimes\Upsilon^{b_{m+1}}).(f\otimes g)(x^{[m+1]})$,

where a_j and b_j are nonnegative integers for each $j=1,...,m+1$, $(A_1\otimes B_1)...(A_k\otimes B_k).(f\otimes g):=(A_1...A_k\otimes B_1...B_k).(f\otimes g):=(A_1...A_kf)(B_1...B_kg)$.

2. Note. Consider the projection

(1) $\psi_n: X^{m(n)}\times\mathbf{K}^{s(n)}\to X^{l(n)}\times\mathbf{K}^n$,

where $m(n)=2m(n-1)$, $s(n)=2s(n-1)+1$, $l(n)=n+1$ for each $n\in\mathbf{N}$ such that $m(0)=1$, $s(0)=0$, $m(n)=2^n$, $s(n)=1+2+2^2+...+2^{n-1}=2^n-1$. Then $m(n)$, $s(n)$, $l(n)$ and n correspond to number of variables in X, **K** for Υ^n, in X and **K** for $\bar{\Phi}^n$ respectively. Therefore, $\psi(x^{[n]})=x^{(n)}$ and $\psi_n(U^{[n]})=U^{(n)}$ for each $n\in\mathbf{N}$ for suitable ordering of variables. Thus $\bar{\Phi}^n f(x^{(n)})=\hat{\psi}_n\Upsilon^n f(x^{[n]})=f^{[n]}(x^{[n]})|_{W^{(n)}}$, where $\hat{\psi}_n g(y):=g(\psi_n(y))$ for a function g on a subset V in $X^{l(n)}\times\mathbf{K}^n$ for each $y\in\psi_n^{-1}(V)\subset X^{m(n)}\times\mathbf{K}^{s(n)}$, $W^{(n)}=U^{(n)}\times 0$, $0\in X^{m(n)-l(n)}\times\mathbf{K}^{s(n)-n}$ for the corresponding ordering of variables.

3. Corollary. *Let either $f,g\in C^n(U,Y)$, where U is an open subset in X, Y is an algebra over* **K**, *or $f\in C^n(U,\mathbf{K})$ and $g\in C^n(U,Y)$, where Y is a topological vector space over* **K**, *then*

(1) $\bar{\Phi}^n(fg)(x^{(n)})=(\bar{\Phi}\otimes\hat{P}+\hat{\pi}\otimes\bar{\Phi})^n.(f\otimes g)(x^{(n)})$

and $\bar{\Phi}^n(fg)\in C^0(U^{(n)},Y)$. In more details:

(2) $\bar{\Phi}^n(fg)(x^{(n)})=\sum_{0\le a,0\le b,a+b=n}\sum_{j_1<...<j_a;s_1<...<s_b;\{j_1,...,j_a\}\cup\{s_1,...,s_b\}=\{1,...,n\}}$
$\bar{\Phi}^a f(x;v_{j_1},...,v_{j_a};t_{j_1},...,t_{j_a})\bar{\Phi}^b g(x+v_{j_1}t_{j_1}+...+v_{j_a}t_{j_a};v_{s_1},...,v_{s_b};t_{s_1},...,t_{s_b})$.

Proof. The operator $\hat{\psi}_n$ is **K**-linear, since $\hat{\psi}_n(af+bg)(y)=(af+bg)(\psi_n(y))=af(\psi_n(y))+bg(\psi_n(y))$ for each $a,b\in\mathbf{K}$ and functions f,g on a subset V in $X^{l(n)}\times\mathbf{K}^n$ and each $y\in\psi_n^{-1}(V)\subset X^{m(n)}\times\mathbf{K}^{s(n)}$. Mention that the restrictions of $\hat{\pi}_k^{k-1}$ and P_k on $W^{(k)}$ gives $\pi_k^{k-1}(x^{(k)}):=x^{(k-1)}$ and $(P_kg)(x^{(k)}):=g(x^{(k-1)}+v_kt_k)$ in the notation of §1.1. The application of the operator $\hat{\psi}_n$ to both sides of Equation 1(1) gives Equation (1) of this corollary, since $\hat{\psi}_n\Upsilon^n=\bar{\Phi}^n$ for each nonnegative integer n, where $\Upsilon^0=I$ and $\bar{\Phi}^0=I$ and $\hat{\psi}_0=I$ are the unit operators.

4. Lemma. *Let $f_1,...,f_k\in C^{[n]}(U,Y)$, where U is an open subset in X, either Y is an algebra over* **K**, *or $f_1,...,f_{k-1}\in C^{[n]}(U,\mathbf{K})$ and $f_k\in C^{[n]}(U,Y)$, where Y is a topological vector space over* **K**, *then*

(1) $(f_1...f_k)^{[n]}(x^{[n]}) = [\sum_{\alpha=0}^{k-1}\hat{\pi}^{\otimes\alpha}\otimes\Upsilon\otimes\hat{P}^{\otimes(k-\alpha-1)}]^n.(f_1\otimes...\otimes f_k)(x^{[n]})$ *and* $(f_1...f_k)^{[n]}\in C^0(U^{[n]},Y)$*, where*

$\hat{\pi}^{\otimes\alpha}\otimes\Upsilon\otimes\hat{P}^{\otimes(k-\alpha-1)}.(f_1\otimes...\otimes f_k) := (\hat{\pi}(f_1...f_\alpha))(\Upsilon f_{\alpha+1})(\hat{P}(f_{\alpha+2}...f_k))$, *where* $\hat{\pi}^0 := I$, $\hat{P}^0 = I$ *is the unit operator,* $\hat{\pi}f_0 := 1$, $\hat{P}f_{k+1} := 1$ *(see Lemma 1).*

Proof. Consider at first $n = 1$ and apply Formula 1(1) by induction to appearing products of functions, then

(2) $\Upsilon^1(f_1...f_k)(x^{[1]}) = [(\Upsilon^1(f_1...f_{k-1}))(P_1f_k) + (\hat{\pi}^1(f_1...f_{k-1}))(\Upsilon^1 f_k)](x^{[1]}) =$
$[(\Upsilon^1(f_1...f_{k-2}))(P_1f_{k-1})(P_1f_k) + (\hat{\pi}^1(f_1...f_{k-2}))(\Upsilon^1 f_{k-1})(P_1f_k)$
$+(\hat{\pi}^1(f_1...f_{k-1}))(\Upsilon^1 f_k)](x^{[1]}) = ...$
$= (\sum_{\alpha=0}^{k-1}(\hat{\pi}^1)^{\otimes\alpha}\otimes\Upsilon^1\otimes P_1^{\otimes(k-\alpha-1)}).(f_1\otimes...\otimes f_k)$,
where $A^{\otimes\alpha}\otimes B\otimes C^{\otimes(k-\alpha-1)}.(f_1\otimes...\otimes f_k) := (A(f_1...f_\alpha))(Bf_{\alpha+1})(C(f_{\alpha+2}...f_k))$ for operators A, B and C and each nonnegative integer α, where $A^0 := I$, $C^0 = I$ is the unit operator, $Af_0 := 1$, $Cf_{k+1} := 1$, in particular, $A = \hat{\pi}^1$, $B = \Upsilon^1$, $C = P_1$. Thus, acting by induction on both sides by Υ^1 from Formula (2) we get Formula (1) of this lemma, since the product of n terms $\Upsilon^1...\Upsilon^1$ is equal to Υ^n.

5. Corollary. *Let* $f_1,...,f_k\in C^n(U,Y)$*, where* U *is an open subset in* X*, either* Y *is an algebra over* $\mathbf{K}$*, or* $f_1,...,f_{k-1}\in C^n(U,\mathbf{K})$ *and* $f_k\in C^n(U,Y)$*, where* Y *is a topological vector space over* $\mathbf{K}$*, then*

(1) $\bar{\Phi}^n(f_1...f_k)(x^{(n)}) = [\sum_{\alpha=0}^{k-1}\hat{\pi}^{\otimes\alpha}\otimes\bar{\Phi}\otimes\hat{P}^{\otimes(k-\alpha-1)}]^n.(f_1\otimes...\otimes f_k)(x^{(n)})$ *and* $\Phi^n(f_1...f_k)\in C^0(U^{(n)},Y)$*, where*

$\hat{\pi}^{\otimes\alpha}\otimes\bar{\Phi}\otimes\hat{P}^{\otimes(k-\alpha-1)}.(f_1\otimes...\otimes f_k) := (\hat{\pi}(f_1...f_\alpha))(\bar{\Phi}f_{\alpha+1})(\hat{P}(f_{\alpha+2}...f_k))$ *(see Lemma 3).*

Proof. Applying operator $\hat{\psi}_n$ from Note 2 to both sides of Equation 4(1) we get Formula (1) of this Corollary.

6. Lemma. *Let* $u\in C^{[n]}(\mathbf{K}^s,\mathbf{K}^m)$*,* $u(\mathbf{K}^s)\subset U$ *and* $f\in C^{[n]}(U,Y)$*, where* U *is an open subset in* $\mathbf{K}^m$*,* $s,m\in\mathbf{N}$*,* Y *is a* $\mathbf{K}$*-linear space, then*

(1) $(f\circ u)^{[n]}(x^{[n]}) = [\sum_{j_1=1}^m...\sum_{j_n=1}^{m(n)}(A_{j_n,v^{[n-1]},t_n}...A_{j_1,v^{[0]},t_1}f\circ u)(\Upsilon^1\circ p_{j_n}\hat{S}_{j_{n-1}+1,v^{[n-2]}t_{n-1}}$
$...\hat{S}_{j_1+1,v^{[0]}t_1}u^{n-1})(P_n\Upsilon^1\circ p_{j_{n-1}}\hat{S}_{j_{n-2}+1,v^{[n-3]}t_{n-2}}...\hat{S}_{j_1+1,v^{[0]}t_1}u^{n-2})...(P_n...P_2\Upsilon^1\circ p_{j_1}u)$
$+\sum_{j_1=1}^m...\sum_{j_{n-1}=1}^{m(n-1)}(\hat{\pi}^1(A_{j_{n-1},v^{[n-2]},t_{n-1}}...A_{j_1,v^{[0]},t_1}f\circ u)[\sum_{\alpha=0}^{n-2}\hat{\pi}^{\otimes\alpha}\otimes\Upsilon\otimes\hat{P}^{\otimes(n-\alpha-2)}]$
$((\Upsilon^1\circ p_{j_{n-1}}\hat{S}_{j_{n-2}+1,v^{[n-3]}t_{n-2}}...\hat{S}_{j_1+1,v^{[0]}t_1}u^{n-2})\otimes...\otimes(P_{n-1}...P_2\Upsilon^1\circ p_{j_1}u))$
$+[\sum_{\alpha=0}^{n-2}\hat{\pi}^{\otimes\alpha}\otimes\Upsilon\otimes\hat{P}^{\otimes(n-\alpha-2)}](\sum_{j_1=1}^m...\sum_{j_{n-2}=1}^{m(n-2)}(\hat{\pi}^1(A_{j_{n-2},v^{[n-3]},t_{n-2}}...A_{j_1,v^{[0]},t_1}f\circ$

$u)) \otimes [\sum_{\alpha=0}^{n-3} \hat{\pi}^{\otimes\alpha} \otimes \Upsilon \otimes \hat{P}^{\otimes(n-\alpha-3)}]((\Upsilon^1 \circ p_{j_{n-2}} \hat{S}_{j_{n-3}+1,v^{[n-4]}t_{n-3}} \dots \hat{S}_{j_1+1,v^{[0]}t_1} u^{n-3}) \otimes$
$\dots \otimes (P_{n-2} \dots P_2 \Upsilon^1 \circ p_{j_1} u)) + \dots$
$+[\sum_{\alpha=0}^{2} \hat{\pi}^{\otimes\alpha} \otimes \Upsilon \otimes \hat{P}^{\otimes(2-\alpha)}]^{n-3} \{\sum_{j_1=1}^{m} \sum_{j_2=1}^{m(2)} (\hat{\pi}^1 A_{j_2,v^{[1]},t_2} A_{j_1,v^{[0]},t_1} f \circ u)(\Upsilon^1 \otimes$
$\hat{P}^1 + \hat{\pi}^1 \otimes \Upsilon^1)((\Upsilon^1 \circ p_{j_2} \hat{S}_{j_1+1,v^{[0]}t_1} u) \otimes (P_2 \Upsilon^1 \circ p_{j_1} u))\}$
$+(\Upsilon \otimes \hat{P} + \hat{\pi} \otimes \Upsilon)^{n-2} \{\sum_{j_1=1}^{m} (\hat{\pi}^1 A_{j_1,v^{[0]},t_1} f \circ u) \otimes (\Upsilon^2 \circ p_{j_1} u)\}](x^{[n]})$
and $f \circ u \in C^0((\mathbf{K}^s)^{[n]}, Y)$, *where*

$S_{j,\tau} u(y) := (u_1(y), \dots, u_{j-1}(y), u_j(y + \tau_{(s)}), u_{j+1}(y + \tau_{(s)}), \dots, u_m(y + \tau_{(s)}))$, $u = (u_1, \dots, u_m)$, $u_j \in \mathbf{K}$ *for each* $j = 1, \dots, m$, $y \in \mathbf{K}^s$, $\tau = (\tau_1, \dots, \tau_k) \in \mathbf{K}^k$, $k \geq s$, $\tau_{(s)} := (\tau_1, \dots, \tau_s)$, $p_j(x) := x_j$, $x = (x_1, \dots, x_m)$, $x_j \in \mathbf{K}$ *for each* $j = 1, \dots, m$, $\hat{S}_{j+1,\tau} g(u(y), \beta) := g(S_{j+1,\tau} u(y), \beta)$, $y \in \mathbf{K}^s$, β *is some parameter,* $A_{j,v,t} := (\hat{S}_{j+1,vt} \otimes t\Upsilon^1 \circ p_j)^* \Upsilon_j$, *where* Υ^1 *is taken for variables* (x, v, t) *or corresponding to them after actions of preceding operations as* Υ^k, $\Upsilon_j^1 f(x, v_j, t) := [f(x + e_j v_j t) - f(x)]/t$, $(B \otimes A)^* \Upsilon^1 f_i \circ u^i(x, v, t) := \Upsilon_j^1 f_i(Bu^i, v, Au^i)$, $B : \mathbf{K}^{m(i)} \to \mathbf{K}^{m(i)}$, $A : \mathbf{K}^{m(i)} \to \mathbf{K}$, $e_j = (0, \dots, 0, 1, 0, \dots, 0) \in \mathbf{K}^{m(i)}$ *with* 1 *on j-th place;* $m(i) = m + i - 1$, $j_i = 1, \dots, m(i)$; $u^1 := u$, $u^2 := (u^1, t_1 \Upsilon^1 \circ p_{j_1} u^1), \dots, u^n = (u^{n-1}, t_{n-1} \Upsilon^1 \circ p_{j_{n-1}} u^{n-1})$, $A_{j_1,v^{[0]},t_1} f \circ u =: f_1 \circ u^1$, $A_{j_n,v^{[n-1]},t_n} f_{n-1} \circ u^{n-1} =: f_n \circ u^n$, $\hat{S}_* \Upsilon^1 f(z) := \Upsilon^1 f(\hat{S}_* z)$.

Proof. At first consider $n = 1$, then $(f \circ u)^{[1]}(t_0, v, t) = [f(u(t_0 + vt)) - f(u(t_0))]/t$, where $t_0 \in \mathbf{K}^s$, $t \in \mathbf{K}$, $v \in \mathbf{K}^s$. Though we consider here the general case mention, that in the particular case $s = 1$ one has $t_0 \in \mathbf{K}$, $v \in \mathbf{K}$. Then

$(f \circ u)^{[1]}(t_0, v, t) = [f(u(t_0 + vt)) - f(u_1(t_0), u_2(t_0 + vt), \dots, u_m(t_0 + vt))]/t + [f(u_1(t_0), u_2(t_0 + vt), u_3(t_0 - vt), \dots, u_m(t_0 + vt)) - f(u_1(t_0), u_2(t_0), u_3(t_0 + vt), \dots, u_m(t_0 + vt))]/t +$
$\dots + [f(u_1(t_0), \dots, u_{m-1}(t_0), u_m(t_0 + vt)) - f(u(t_0))]/t$,
where $u = (u_1, \dots, u_m)$, $u_j \in \mathbf{K}$ for each $j = 1, \dots, m$. Since $u_j(t_0 + vt) - u_j(t_0) = tu_j^{[1]}(t_0, v, t)$, hence

$(f \circ u)^{[1]}(t_0, v, t) = \Upsilon^1 f((u_1(t_0), u_2(t_0 + vt), \dots, u_m(t_0 + vt)), e_1, t\Upsilon^1 u_1(t_0, v, t)) \Upsilon^1 u_1(t_0, v, t) + \Upsilon^1 f((u_1(t_0), u_2(t_0), u_3(t_0 + vt), \dots, u_m(t_0 + vt)), e_2, t\Upsilon^1 u_2(t_0, v, t)) \Upsilon^1 u_2(t_0, v, t) + \dots + \Upsilon^1 f(u(t_0), e_m, t\Upsilon^1 u_m(t_0, v, t)) \Upsilon^1 u_m(t_0, v, t)$,
since $u_j \in \mathbf{K}$ for each $j = 1, \dots, m$ and $\mathbf{K}$ is the field, where $e_j = (0, \dots, 0, 1, 0, \dots, 0) \in \mathbf{K}^m$ with 1 on j-th place for each $j = 1, \dots, m$. With the help of shift operators it is possible to write the latter formula shorter:

(2) $\Upsilon^1(f \circ u)(y, v, t) = \sum_{j=1}^{m} \hat{S}_{j+1,vt} \Upsilon^1 f(u(y), e_j, t\Upsilon^1 \circ p_j u(y, v, t))(\Upsilon^1 \circ p_j u(y, v, t))$,
where $p_j(x) := x_j$, $x = (x_1, \dots, x_m)$, $x_j \in \mathbf{K}$ for each $j = 1, \dots, m$,

$\hat{S}_{j+1,\tau}g(u(y),\beta) := g(S_{j+1,\tau}u(y),\beta)$, $y \in \mathbf{K}^s$, $\tau \in \mathbf{K}^k$, $k \geq s$, β is some parameter. Introduce operators $A_{j,v,t} := (\hat{S}_{j+1,vt} \otimes t\Upsilon^1 \circ p_j)^* \Upsilon^1_j$, where Υ^1 is taken for variables (y,v,t) or corresponding to them after actions of preceding operators as Υ^k remembering that $y^{[k]}, v^{[k]} \in (\mathbf{K}^s)^{[k]}$, $t \in \mathbf{K}$, $v^{[k]} = (v_1^{[k]}, v_2^{[k]}, v_3^{[k]})$ with $v_1^{[k]}, v_2^{[k]} \in (\mathbf{K}^s)^{[k-1]}$, $v_3^{[k]} \in \mathbf{K}^k$ for each $k \geq 1$, in particular, $v^{[0]} = v_1^{[0]}$ for $k = 0$, $\Upsilon^1_j f(x,v,t) := [f(x+e_jv_jt) - f(x)]/t$, $(B \otimes A)^* \Upsilon^1 f_i \circ u^i(y,v,t) := \Upsilon^1_j f_i(Bu^i, v, Au^i)$, $B : \mathbf{K}^{m(i)} \to \mathbf{K}^{m(i)}$, $A : \mathbf{K}^{m(i)} \to \mathbf{K}$. For example, in the particular case of $s = 1$ we have $v^{[k]} \in (\mathbf{K})^{[k]}$. Therefore, in the general case Formula (2) takes the form:

(3) $\Upsilon^1 f \circ u(y,v,t) = \sum_{j=1}^m (A_{j,v,t} f \circ u)(\Upsilon^1 \circ p_j u)(y,v,t)$.

Take now $n = 2$, then

$\Upsilon^2 f \circ u(y^{[2]}) = \Upsilon^1 \sum_{j=1}^m [(A_{j,v,t} f \circ u)(\Upsilon^1 \circ p_j u)(y,v,t)](y^{[2]})$.

In the square brackets there is the product, hence from Formula 1(1) and Lemma 2.3 we get:

(4) $\Upsilon^2 f \circ u(y^{[2]}) = \sum_{j=1}^m [(\Upsilon^1 A_{j,v^{[0]},t} f \circ u)(P_2 \Upsilon^1 \circ p_j u) + (\hat{\pi}^1 A_{j,v^{[0]},t} f \circ u)(\Upsilon^2 \circ p_j u)](y^{[2]})$.

Then from Formula (3) applied to terms $A_{j,v,t} f \circ u$ it follows, that $\Upsilon^1 A_{j_1,v^{[0]},t_1} f \circ u(y^{[2]}) = \sum_{j_2=1}^{m(2)} (A_{j_2,v^{[1]},t_2} A_{j_1,v^{[0]},t_1} f \circ u)(\Upsilon^1 \circ p_{j_2} S_{j_1+1,v^{[0]}t_1} u)(y^{[2]})$, where $v^{[0]} = v$, $t_1 = t$ (see also Lemma 1). Therefore,

(5) $\Upsilon^2 f \circ u(y^{[2]}) = [\sum_{j_1=1}^m \sum_{j_2=1}^{m(2)} (A_{j_2,v^{[1]},t_2} A_{j_1,v^{[0]},t_1} f \circ u)(\Upsilon^1 \circ p_{j_2} \hat{S}_{j_1+1,v^{[0]}t_1} u)(P_2 \Upsilon^1 \circ p_{j_1} u) + \sum_{j_1=1}^m (\hat{\pi}^1 A_{j_1,v^{[0]},t_1} f \circ u)(\Upsilon^2 \circ p_{j_1} u)](y^{[2]})$.

Then for $n = 3$ applying Formulas (3) and 4(1) to (5) we get:

(6) $\Upsilon^3 f \circ u(y^{[3]}) = [\sum_{j_1=1}^m \sum_{j_2=1}^{m(2)} \sum_{j_3=1}^{m(3)} (A_{j_3,v^{[2]},t_3} A_{j_2,v^{[1]},t_2} A_{j_1,v^{[0]},t_1} f \circ u)$
$(\Upsilon^1 \circ p_{j_3} \hat{S}_{j_2+1,v^{[1]}t_2} \hat{S}_{j_1+1,v^{[0]}t_1} u^2)(P_2 \Upsilon^1 \circ p_{j_2} \hat{S}_{j_1+1,v^{[0]}t_1} u)(P_3 P_2 \Upsilon^1 \circ p_{j_1} u) +$
$\sum_{j_1=1}^m \sum_{j_2=1}^{m(2)} [(\hat{\pi}^1 (A_{j_2,v^{[1]},t_2} A_{j_1,v^{[0]},t_1} f \circ u))(\Upsilon^2 \circ p_{j_2} \hat{S}_{j_1+1,v^{[0]}t_1} u)(P_3 P_2 \Upsilon^1 \circ p_{j_1} u) +$
$(\hat{\pi}^1 \{(A_{j_2,v^{[1]},t_2} A_{j_1,v^{[0]},t_1} f \circ u)(\Upsilon^1 \circ p_{j_2} \hat{S}_{j_1+1,v^{[0]}t_1} u)\}(\Upsilon^1 P_2 \Upsilon^1 \circ p_{j_1} u)] +$
$\sum_{j_1=1}^m \sum_{j_3=1}^{m(3)} (A_{j_3,v^{[2]},t_3} \hat{\pi}^1 A_{j_1,v^{[0]},t_1} f \circ u)(\Upsilon^1 \circ p_{j_3} \hat{S}_{j_1+1,v^{[0]}t_1} u)(P_3 \Upsilon^2 \circ p_{j_1} u)$
$+ \sum_{j_1=1}^m (\hat{\pi}^2 A_{j_1,v^{[0]},t_1} f \circ u)(\Upsilon^3 \circ p_{j_1} u)](y^{[3]})$.

Thus Formula (1) is proved for $n = 1,2,3$. Suppose that it is true for $k = 1,\dots,n$ and prove it for $k = n+1$. Applying Formula 4(1) to both sides of (1) we get:

(7) $\Upsilon^{n+1} f \circ u(y^{[n+1]}) = [\sum_{j_1=1}^m \dots \sum_{j_{n+1}=1}^{m(n+1)} (A_{j_{n+1},v^{[n]},t_{n+1}} \dots A_{j_1,v^{[0]},t_1} f \circ u)$
$(\Upsilon^1 \circ p_{j_{n+1}} \hat{S}_{j_n+1,v^{[n-1]}t_n} \dots \hat{S}_{j_1+1,v^{[0]}t_1} u^n)(P_{n+1} \Upsilon^1 \circ p_{j_n} \hat{S}_{j_{n-1}+1,v^{[n-2]}t_{n-1}} \dots \hat{S}_{j_1+1,v^{[0]}t_1} u^{n-1}) \dots$
$(P_{n+1} \dots P_2 \Upsilon^1 \circ p_{j_1} u) + \sum_{j_1=1}^m \dots \sum_{j_n=1}^{m(n)} (\hat{\pi}^1 (A_{j_n,v^{[n-1]},t_n} \dots A_{j_1,v^{[0]},t_1} f \circ u)$

$\Upsilon^1((\Upsilon^1 \circ p_{j_n}\hat{S}_{j_{n-1}+1,v^{[n-2]}t_{n-1}}...\hat{S}_{j_1+1,v^{[0]}t_1}u^{n-1})...(P_n...P_2\Upsilon^1 \circ p_{j_1}u))+$
$\Upsilon^1(\sum_{j_1=1}^{m}...\sum_{j_{n-1}=1}^{m(n-1)}(\hat{\pi}^1(A_{j_{n-1},v^{[n-2]},t_{n-1}}...A_{j_1,v^{[0]},t_1}f \circ u))\Upsilon^1((\Upsilon^1 \circ p_{j_{n-1}}\hat{S}_{j_{n-2}+1,v^{[n-3]}t_{n-2}}$
$...\hat{S}_{j_1+1,v^{[0]}t_1}u^{n-2})...(P_{n-1}...P_2\Upsilon^1 \circ p_{j_1}u)) + ... + \Upsilon^{n-2}\{\sum_{j_1=1}^{m}\sum_{j_2=1}^{m(2)}$
$(\hat{\pi}^1 A_{j_2,v^{[1]},t_2}A_{j_1,v^{[0]},t_1}f \circ u)\Upsilon^1((\Upsilon^1 \circ p_{j_2}\hat{S}_{j_1+1,v^{[0]}t_1}u)(P_2\Upsilon^1 \circ p_{j_1}u))\}+$
$\Upsilon^{n-1}\{\sum_{j_1=1}^{m}\hat{\pi}^1 A_{j_1,v^{[0]},t_1}f \circ u)(\Upsilon^2 \circ p_{j_1}u)\}](y^{[n+1]}) =$
$[\sum_{j_1=1}^{m}...\sum_{j_{n+1}=1}^{m(n+1)}(A_{j_{n+1},v^{[n]},t_{n+1}}...A_{j_1,v^{[0]},t_1}f \circ u)(\Upsilon^1 \circ p_{j_{n+1}}\hat{S}_{j_n+1,v^{[n-1]}t_n}$
$...\hat{S}_{j_1+1,v^{[0]}t_1}u^n)(P_{n+1}\Upsilon^1 \circ p_{j_n}\hat{S}_{j_{n-1}+1,v^{[n-2]}t_{n-1}}...\hat{S}_{j_1+1,v^{[0]}t_1}u^{n-1})...(P_{n+1}...P_2\Upsilon^1 \circ p_{j_1}u)$
$+\sum_{j_1=1}^{m}...\sum_{j_n=1}^{m(n)}(\hat{\pi}^1(A_{j_n,v^{[n-1]},t_n}...A_{j_1,v^{[0]},t_1}f \circ u)[\sum_{\alpha=0}^{n-1}\hat{\pi}^{\otimes\alpha}\otimes\Upsilon\otimes\hat{P}^{\otimes(n-\alpha-1)}]$
$((\Upsilon^1 \circ p_{j_n}\hat{S}_{j_{n-1}+1,v^{[n-2]}t_{n-1}}...\hat{S}_{j_1+1,v^{[0]}t_1}u^{n-1})\otimes...\otimes(P_n...P_2\Upsilon^1 \circ p_{j_1}u))$
$+[\sum_{\alpha=0}^{n-1}\hat{\pi}^{\otimes\alpha}\otimes\Upsilon\otimes\hat{P}^{\otimes(n-\alpha-1)}](\sum_{j_1=1}^{m}...\sum_{j_{n-1}=1}^{m(n-1)}(\hat{\pi}^1(A_{j_{n-1},v^{[n-2]},t_{n-1}}...A_{j_1,v^{[0]},t_1}f \circ u))\otimes[\sum_{\alpha=0}^{n-2}\hat{\pi}^{\otimes\alpha}\otimes\Upsilon\otimes\hat{P}^{\otimes(n-\alpha-2)}]$
$((\Upsilon^1 \circ p_{j_{n-1}}\hat{S}_{j_{n-2}+1,v^{[n-3]}t_{n-2}}...\hat{S}_{j_1+1,v^{[0]}t_1}u^{n-2})\otimes...\otimes(P_{n-1}...P_2\Upsilon^1 \circ p_{j_1}u))$
$+[\sum_{\alpha=0}^{2}\hat{\pi}^{\otimes\alpha}\otimes\Upsilon\otimes\hat{P}^{\otimes(2-\alpha)}]^{n-2}\{\sum_{j_1=1}^{m}\sum_{j_2=1}^{m(2)}(\hat{\pi}^1 A_{j_2,v^{[1]},t_2}A_{j_1,v^{[0]},t_1}f \circ u)(\Upsilon^1 \otimes \hat{P}^1 + \hat{\pi}^1 \otimes \Upsilon^1)((\Upsilon^1 \circ p_{j_2}\hat{S}_{j_1+1,v^{[0]}t_1}u)\otimes(P_2\Upsilon^1 \circ p_{j_1}u))\}$
$+(\Upsilon\otimes\hat{P}+\hat{\pi}\otimes\Upsilon)^{n-1}\{\sum_{j_1=1}^{m}(\hat{\pi}^1 A_{j_1,v^{[0]},t_1}f \circ u)\otimes(\Upsilon^2 \circ p_{j_1}u)\}](y^{[n+1]})$.

Mention that in general $(\Upsilon^{n-1}f \circ u)(y^{[n+1]})$ may depend nontrivially on all components of the vector $y^{[n-1]}$ through several terms in Formula (7). Thus Formula (1) of this Lemma is proved by induction.

7. Corollary. *Let $u \in C^n(\mathbf{K}^s,\mathbf{K}^m)$, $u(\mathbf{K}^s) \subset U$ and $f \in C^n(U,Y)$, where U is an open subset in $\mathbf{K}^m$, $s,m \in \mathbf{N}$, Y is a $\mathbf{K}$-linear space, then*

(1) $\bar{\Phi}^n(f \circ u)(x^{(n)}) = [\sum_{j_1=1}^{m}...\sum_{j_n=1}^{m(n)}(B_{j_n,v^{(n-1)},t_n}...B_{j_1,v^{(0)},t_1}f \circ u)$
$(\bar{\Phi}^1 \circ p_{j_n}\hat{S}_{j_{n-1}+1,v^{(n-2)}t_{n-1}}...\hat{S}_{j_1+1,v^{(0)}t_1}u^{n-1})(P_n\bar{\Phi}^1 \circ p_{j_{n-1}}\hat{S}_{j_{n-2}+1,v_0^{(n-3)},t_{n-2}}...\hat{S}_{j_1+1,v^{(0)}t_1}u^{n-2})$
$...(P_n...P_2\bar{\Phi}^1 \circ p_{j_1}u) + \sum_{j_1=1}^{m}...\sum_{j_{n-1}=1}^{m(n-1)}(\hat{\pi}^1(B_{j_{n-1},v^{(n-2)},t_{n-1}}...B_{j_1,v^{(0)},t_1}f \circ u)[\sum_{\alpha=0}^{n-2}\hat{\pi}^{\otimes\alpha}\otimes\bar{\Phi}\otimes\hat{P}^{\otimes(n-\alpha-2)}]$
$((\bar{\Phi}^1 \circ p_{j_{n-1}}\hat{S}_{j_{n-2}+1,v^{(n-3)}t_{n-2}}...\hat{S}_{j_1+1,v^{(0)}t_1}u^{n-2})\otimes...\otimes(P_{n-1}...P_2\bar{\Phi}^1 \circ p_{j_1}u))$
$+[\sum_{\alpha=0}^{n-2}\hat{\pi}^{\otimes\alpha}\otimes\bar{\Phi}\otimes\hat{P}^{\otimes(n-\alpha-2)}](\sum_{j_1=1}^{m}...\sum_{j_{n-2}=1}^{m(n-2)}(\hat{\pi}^1(B_{j_{n-2},v^{(n-3)},t_{n-2}}...B_{j_1,v^{(0)},t_1}f \circ u))\otimes[\sum_{\alpha=0}^{n-3}\hat{\pi}^{\otimes\alpha}\otimes\bar{\Phi}\otimes\hat{P}^{\otimes(n-\alpha-3)}]((\bar{\Phi}^1 \circ p_{j_{n-2}}\hat{S}_{j_{n-3}+1,v^{(n-4)}t_{n-3}}...\hat{S}_{j_1+1,v^{(0)}t_1}u^{n-3})\otimes$
$...\otimes(P_{n-2}...P_2\bar{\Phi}^1 \circ p_{j_1}u)) + ...$

$+[\sum_{\alpha=0}^{2}\hat{\pi}^{\otimes\alpha}\otimes\bar{\Phi}\otimes\hat{P}^{\otimes(2-\alpha)}]^{n-3}\{\sum_{j_1=1}^{m}\sum_{j_2=1}^{m(2)}(\hat{\pi}^1 B_{j_2,v^{(1)},t_2}B_{j_1,v^{(0)},t_1}f\circ u)(\bar{\Phi}^1\otimes \hat{P}^1+\hat{\pi}^1\otimes\bar{\Phi}^1)((\bar{\Phi}^1\circ p_{j_2}\hat{S}_{j_1+1,v^{(0)}t_1}u)\otimes(P_2\bar{\Phi}^1\circ p_{j_1}u))\}$
$+(\bar{\Phi}\otimes\hat{P}+\hat{\pi}\otimes\bar{\Phi})^{n-2}\{\sum_{j_1=1}^{m}(\hat{\pi}^1 B_{j_1,v^{(0)},t_1}f\circ u)\otimes(\bar{\Phi}^2\circ p_{j_1}u)\}](x^{(n)})$
and $f\circ u\in C^0((\mathbf{K}^s)^{(n)},Y)$ *(see notation of Lemma 9), where* $B_{j,v,t}:=(\hat{S}_{j+1,vt}\otimes t\bar{\Phi}^1\circ p_j)^*\bar{\Phi}^1_j$, *where* $\bar{\Phi}^1$ *is taken for variables* (x,v,t) *or corresponding to them after actions of preceding operations as* $\bar{\Phi}^k$, $\bar{\Phi}^1_j f(x,v,t):=[f(x+e_jv_jt)-f(x)]/t$, $(B\otimes A)^*\bar{\Phi}^1 f_i\circ u^i(x,v,t):=\bar{\Phi}^1_j f_i(Bu^i,v,Au^i)$, $B:\mathbf{K}^{m(i)}\to\mathbf{K}^{m(i)}$, $A:\mathbf{K}^{m(i)}\to\mathbf{K}$, $m(i)=m+i-1$, $j_i=1,\dots,m(i)$, $u^1=u$, $u^2:=(u^1,t_1\bar{\Phi}^1\circ p_{j_1}u^1)$, $u^n:=(u^{n-1},t_{n-1}\bar{\Phi}^1\circ p_{j_{n-1}}u^{n-1})$, $\hat{S}_*\bar{\Phi}^1 f(x):=\bar{\Phi}^1 f(\hat{S}_*x)$.

Proof. The restriction of operators of Lemma 6 on $W^{(n)}$ from Note 2 gives Formula (1) of this corollary, where $v^{(k)}\in(\mathbf{K}^s)^k\times\mathbf{K}^k$.

References

[1] Y. Amice. "Interpolation p-Adique". *Bull. Soc. Math. France* **92**(1964), 117-180.

[2] Aref'eva I.Ya., Dragovich B., Frampton P.H., Volovich I.V. "Wave functions of the universe and p-adic gravity". *Int. J. Modern Phys.* **6** (1991), 4341-4358.

[3] J. Araujo, W.H. Schikhof. "The Weierstrass-Stone approximation theorem for p-adic C^n-functions". *Ann. Math. Blaise Pascal.* **1** (1994), 61-74.

[4] Yu.A. Bachturin. "Basic structures of modern algebra" (Moscow: Nauka, 1990).

[5] A. Banyaga. "The structure of classical diffeomorphism groups". *Mathem. and its Applic.* **400** (Dordrecht: Kluwer, 1997

[6] N. Bourbaki. "Variétés différentielles et analytiques". *Fasc. XXXIII* (Paris: Hermann, 1967).

[7] P.-J. Cahen, J.-L. Chabert. "On the ultrametric Stone-Weierstrass theorem on Mahler's expansion". *J. de Théorie des Nombres de Bordeaux* **14** (2002), 43-57.

[8] S. De Smedt. "Local invertibility of non-archimedean vector-valued functions". *Ann. Math. Blaise Pascal* **5** (1998), 13-23.

[9] B. Diarra. "Ultraproduits ultrametriques de corps values". *Ann. Sci. Univ. Clermont II, Sér. Math.* **22** (1984), 1-37.

[10] B. Diarra. "On reducibility of ultrametric almost periodic linear representations". *Glasgow Math. J.* **37** (1995), 83-98.

[11] R. Engelking. "General topology". *2-nd ed., Sigma Series in Pure Mathematics,* V. 6 (Berlin: Heldermann Verlag, 1989).

[12] J.M.G. Fell, R.S. Doran. "Representations of $*$-algebras, locally compact groups, and Banach $*$-algebraic bundles". V. 1 and V. 2 (Boston: Acad. Press, 1988).

[13] S. Haller, J. Teichmann. "Smooth perfectness through decomposition of diffeomorphisms into fiber preserving ones". *Annals of Global Analysis and Geometry* **23** (2003), 53-63.

[14] K. Hensel. *Jahresber. Deutsch. Math. Ver.* **6: 1** (1899), 83-88.

[15] I.N. Herstein. "Topics in algebra" (New-York: Blaisdell Publ. Co., 1964).

[16] E. Hewitt, K.A. Ross. "Abstract harmonic analysis" (Berlin: Springer, 1979).

[17] Y. Jang. "Non-Archimedean quantum mechanics". *Tohoku Math. Publ.* **10** (1998).

[18] I. Kaplansky. "The Weiertsrass theorem in fields with valuations". *Proc. Amer. Math. Soc.* **1** (1950), 356-357.

[19] M.I. Kargapolov, Yu.I. Merzljakov. "Principles of group theory" (Moscow: Nauka, 1977).

[20] W. Ling. "Factorizable groups of homeomorphisms". *Compositio Mathem.* **51** (1984), 41-50.

[21] S.V. Ludkovsky. "Irreducible unitary representations of non-Archimedean groups of diffeomorphisms". *Southeast Asian Bull. of Mathem.* **22: 3** (1998), 419-436.

[22] S.V. Ludkovsky. "Measures on groups of diffeomorphisms of non-Archimedean manifolds, representations of groups and their applications". *Theoret. i Math. Phys.* **119: 3** (1999), 381-396.

[23] S.V. Ludkovsky. "Quasi-invaraint measures on non-Archimedean groups and semigroups of loops and paths, their representations. I, II". *Ann. Math. Blaise Pascal* **7: 2** (2000), 19-53, 55-80.

[24] S.V. Ludkovsky. "A structure and representations of diffeomorphism groups of non-archimedean manifolds". *Southeast Asian Bull. of Mathem.* **26** (2003), 975-1004 (previous variant: *Los Alamos Nat. Lab.* **math.GR/0004126**, 32 pages, 19 April 2000).

[25] S.V. Ludkovsky, B. Diarra. "Profinite and finite groups associated with loop and diffeomorphism groups of non-archimedean manifolds". *Int. J. of Math. and Mathematical Sciences* **2003: 42** (2003), 2673-2688.

[26] S.V. Ludkovsky. "Quasi-invariant and pseudo-differentiable measures on non-archimedean Banach spaces with values in non-archimedean fields". *J. Math. Sci.* **122: 1** (2004), 2949-2983 (previous variant: Los Alamos Nat. Lab. preprint **math.GM/0106170**).

[27] S.V. Ludkovsky. "Non-archimedean polyhedral expansions of ultrauniform spaces". *Russ. Math. Surveys. V.* **54: 5** (1999), 163-164. (in detail: Los Alamos National Laboratory, USA. Preprint **math.AT/0005205**, 39 pages, 22 May 2000).

[28] S.V. Ludkovsky. "Non-archimedean polyhedral decompositions of ultrauniform spaces". *Fund. Prikl. Mathem.* **6: 2** (2000), 455-475.

[29] S.V. Ludkovsky. "Embeddings of non-archimedean Banach manifolds into non-archimedean Banach spaces". *Russ. Math. Surv.* **53** (1998), 1097-1098.

[30] S.V. Ludkovsky. "Differentiability of functions: approximate, global and differentiability along curves over non-archimedean fields". *J. Mathem. Sci. [transl. from: Contemporary Mathem. and its Applic., V. 52, Functional Anal.]* **157**: 2 (2009), 311-366 (previous variant: Los Alamos Nat. Lab. math.CA/0608724).

[31] P.W. Michor. "Manifolds of differentiable mappings" (Orpington: Shiva Publishing Limited, 1980).

[32] L. Narici, E. Beckenstein. "Topological vector spaces" (New York: Marcel Dekker Inc., 1985).

[33] A. Ostrowski. "Untersuchungen zur arithmetischen Theorie der Körper". *Math. Zeit.* **39** (1935), 269-404.

[34] "p-Adic Mathematical Physics." 2nd International Conference (Belgrad, 2005). Editors: A.Yu. Khrennikov, Z. Rakić, I.V. Volovich. *AIP Conference Proceedings* **826**, New York, 2006.

[35] L.S. Pontrjagin. "Continuous groups" (Moscow: Nauka, 1984).

[36] J. Riordan. "Combinatorial identities" (New York: John Wiley and Sons Inc., 1968).

[37] A. Robert. "Representations p-adiques irréductibles de sous-groupes ouverts de $SL_2(\mathbf{Z_p})$". *C.R. Acad. Sci. Paris Sér. I Math.* **298: 11** (1984), 237-240.

[38] A.C.M. van Rooij. "Non-Archimedean functional analysis" (New York: Marcel Dekker Inc., 1978).

[39] A.C.M. van Rooij, W.H. Schikhof. "Groups representations in non-Archimedean Banach spaces". *Bull. Soc. Math. France. Memoire.* **39-40** (1974), 329-340.

[40] W.H. Schikhof. "Ultrametric calculus" (Cambridge: Cambr. Univ. Press, 1984).

[41] W.H. Schikhof. "Non-Archimedean calculus". *Nijmegen: Math. Inst., Cath. Univ., Report* **7812**, 130 pages, 1978.

[42] V.S. Vladimirov, I.V. Volovich, E.I. Zelenov. "p-adic analysis and mathematical physics" (Moscow: Nauka, 1994).

[43] A. Weil. "Basic number theory" (Berlin: Springer, 1973).

In: Mathematical Theories, Models and Groups ISBN 978-1-61324-250-6
Editor: Marcus J. Grecho

Chapter 7

THE CONCEPT OF A SUITABLE INSURANCE POLICY USING LEADER-FOLLOWER GAMES

***Ilaria Colivicchi*[1], *Fausto Mignanego*[2] *and Sabrina Mulinacci*[3]**
[1]Department of Mathematics for Decisions
University of Florence, Florence, Italy
[2]Institute of Econometric and Mathematics
for Economic, Financial and Actuarial Decisions
Catholic University of Milan, Milan, Italy
[3]Department of Mathematical Economics
University of Bologna, Bologna, Italy

1. Introduction

This paper focuses on a new method to create a suitable new insurance product; the aim is, obviously, to satisfy the needs of the potential policy holders maximizing the expected utility both of the company and the agent, as far as possible. By analyzing the Italian life insurance market, we can observe a new trend for the companies that aim to create policies for different sets of policy owners. Life policy can today be custom-made to suit the client's needs and new unit-linked policies have been conceived precisely to reach a specific segment of the market.

Such a behavior represents a new approach for the companies, which so far have not focused their attention on this problem. This explains why no relevant actuarial literature on this topic exists.

Leader-follower games (see [1, 2, 6, 8, 10, 12] for a complete treatment on this subject) could be a new approach to model the actuarial problem presented above; these games fit the classical definition of Stackelberg games, but in this paper the equilibria we are looking for must be considered, more generally, hierarchic games solutions.

In our model, the leader is the insurance company, which is the first level agent in the game; the follower or second level agent is the individual who has to sell the policy to a particular class of clients. The target each one wants to reach, playing the game, is to maximize their own expected utilities.

The insurance company creates a life insurance product keeping in mind, among the others, the needs of the market, the segment of clients it wants to reach, the relationship to the follower in playing the game.

In particular, the insurance company, which aims to conquer a certain set of clients, proposes different rates to the agent in order to put him in the position of selling the product to the classes of clients which correspond, as close as possible, to the target set.

The follower tries to maximize her expected utility by selling the policy to clients having a profile as similar as possible to the target given by the company. Her behavior depends on one's own utility and on the characteristics of the product.

The game could be repeated in many following time-intervals according to agreements among the players. The follower can choose among some possible strategies and, once the game is played, the expected utility of the leader could be close to the global solution. If both players reach a satisfactory solution, then a hierarchic equilibrium of Stackelberg type is obtained.

Our goal is to show how leader-follower games can help an insurance company to build a life product (in this work a unit-linked policy is considered), if a specific set of clients is defined as a target.

This way, our research represents an innovation in the actuarial life insurance literature, since such a game model has never been applied before to the scope of creating a new life product.

The authors thank the anonymous referees for the comments and the helpful suggestions that have enabled to reshape the paper into its current version. They are also grateful to Professor M. Galeotti for reading carefully the paper and helped them in rephrasing it.

2. Mathematical Tools

This Section is devoted to a brief introduction to the mathematical instruments needed in the sequel to construct the game model by which the interaction between a company and an agent will be simulated.

2.1. Expected Utility Maximization

We assume that each player can value several possible incomes in terms of one's own subjective preferences in risky situations; therefore we will consider expected utilities. The interested reader is referred to [7] for a short presentation (for a more complete one see, for example, [5]).

In our case, as it will be specified in the following, the market of possible policy holders is segmented into classes (i, j), $1 \leq i \leq N$, $1 \leq j \leq M$, so that the sale of policy to clients of one class produces a stochastic income $X_{i,j}$.

Hence, under the von Neumann-Morgenstern axioms, both the company and the agent express their preferences through expected utilities of the type

$$U(X) = \mathbb{E}\left[u(X)\right],$$

where X denotes the $N \times M$ stochastic matrix of the $X_{i,j}$.

Let us remark that, although all the realizations of the $X_{i,j}$ are monetary values, they are not evaluated in the same way by the company, which aims to reach certain segments of the market rather than others, neither consequently by the agent, who receives different incentives according to the typology of her clients.

Practically, in the model we will provide, incomes coming from different classes of clients will be differently "weighed", both from the company and the agent.

2.2. Stackelberg Games

The game we have considered to model the interaction between the company and the agent is a Stackelberg game.

Stackelberg games will be briefly illustrated below (see [1, 2, 6, 8, 10] for a complete treatment on this subject).

Let Ξ and Υ two vector spaces. $u^L : \Xi \times \Upsilon \to \mathbb{R}$ is the utility function of the player called the leader, L, while $u^F : \Xi \times \Upsilon \to \mathbb{R}$ is the utility function of the player called the follower, F.

L announces $z \in \Xi$ to F, who reacts with

$$O(z) = \left\{ \tilde{y} \in \Upsilon : u^F(z,\tilde{y}) \geq u^F(z,y), \text{ for every } y \in \Upsilon \right\} \tag{2.1}$$

(we suppose to be able to guarantee $O(z) \neq \emptyset$ for all z). Let $\hat{y}(z) \in O(z)$ be the particular strategy chosen by F for each $z \in \Xi$.

Hence L optimizes her own utility function as follows:

$$\max_{z \in \Xi} u^L(z, \hat{y}(z)) = u^L(z', y'),$$

where $y' = \hat{y}(z') \in O(z')$ and z' is supposed to exist.

$[z', y']$ is a solution of the Stackelberg problem.

The relationship between L and F is uncooperative and L is rationally uncertain on the behavior of F. In fact, F is not interested in the particular choice of $y \in O(z)$ (which, in general, is not a singleton), while L is interested in a pair (z^*, y^*) that satisfies

$$u^L(z^*, y^*) = \max_{z \in \Xi} \max_{y \in \Upsilon} u^L(z, y).$$

¿From a mathematical point of view (z^*, y^*) is attainable under suitable assumptions, but in a concrete framework (z^*, y^*) could be considered a "utopia-point". Thus, for some fixed $\varepsilon > 0$, any point $(\check{z}, \check{y})$ such that $\check{y} \in O(\check{z})$ and

$$u^L(z^*, y^*) - \varepsilon \leq u^L(\check{z}, \check{y}) \leq u^L(z^*, y^*)$$

is desirable for L.

On the other hand, $\check{y} \in O(\check{z})$ means $U^F(\check{z}, \check{y}) \geq U^F(\check{z}, y)$, for every $y \in \Upsilon$.

In conclusion, $[\check{z}, \check{y}]$ is a hierarchical equilibrium point and it is called "intermediate solution" of the Stackelberg problem (see [2, 8, 9]). This solution is as close as we like to the global maximum since ε is an arbitrarily small positive number.

3. The Game

We are now ready to model the game which would enable the company to force the agent in selling a specific policy to a particular set of clients.

Let us consider a partition of the set of all possible buyers of a particular policy with respect to their jobs and income levels. We connect to each job an integer number. $K = \{1, 2, \ldots, i, \ldots, N\}$ is the set of all considered jobs, in the sense that each integer number is associated to a particular set of potential clients representing the same workers category (see Section 4 as an example). We can divide the set of all income levels into M ranges, numbered by $j = 1, 2, \ldots, M$.

The couple (i, j) represents the class of all the clients that perform the same job i and have the their income level in the range indicated by j.

Let $\mathbb{M}$ be the set of all matrices $N \times M$. Given $A = (a_{i,j}) \in \mathbb{M}$ and $B = (b_{i,j}) \in \mathbb{M}$, we define $A \cdot B$ the matrix $C = (c_{i,j}) \in \mathbb{M}$ such that $c_{i,j} = a_{i,j} b_{i,j}$ (this is not the usual rows times columns product).

Let $\Delta = \{\beta \in \mathbb{M} : \beta_{i,j} \in [0, 1]\}$ be the set of all the participating rates the company can give to the agent, as a function of the buyer's class (the company proposes to the agent a specific $\beta_{i,j}$ as reward for selling the policy to a client in the class (i, j)).

Let $\pi_{i,j}$ be the number of all possible clients visited by the agent in a specific time interval that are in the class (i, j). Let

$$\mathcal{P} = \left\{P = (\pi_{i,j})_{i=1,\ldots,N, j=1,\ldots,M} \in \mathbb{M}\right\}$$

such that $\sum_{i,j} \pi_{i,j} \leq N_A$ where N_A represents the maximum number of people the agent can visit in the fixed time interval.

For any $P \in \mathcal{P}$, $Y^P = \left(Y^P_{i,j}\right)$ is the $N \times M$ random matrix representing the quantities of policies really sold, verifying the condition $Y^P_{i,j} \leq \pi_{i,j}$. Let μ be its probability distribution.

Assuming, by sake of simplicity, the policy yearly premium equal to one, the agent's income is the random variable $W = \sum_{i,j} \beta_{i,j} Y^P_{i,j}$.

Let $U^F_\mu : \Delta \times \mathcal{P} \to \mathbb{R}$ be the agent's expected utility function, defined by

$$U^F_\mu(\beta, P) = \mathbb{E}^\mu\left[u\left(\beta \cdot Y^P\right)\right],$$

with $u : \mathbb{M} \to \mathbb{R}$, where $u(A)$ is an increasing function with respect to each entry of A.

Remark 3.1. *Since it makes sense to suppose that, in some cases, the agent may prefer one class (i, j) to the others, we assume her utility to be a function, separately, of the incomes $\beta_{i,j} Y^P_{i,j}$ obtained from selling the policy to a client in each particular class (that is, her utility function is state dependent, see Subsection 2.1).*

Remark 3.2. *Supposing all the clients have independent behavior, Y^P is a random matrix whose entries are independent and binomially distributed random variables. This way, if $y \in \mathbb{M}$ is one value that Y^P may assume and $\mu(i,j)$ is the probability that the agent sells a policy to a clients in the class (i,j), it holds*

$$\mu\left(Y^P = y\right) = \prod_{i,j} \binom{\pi_{i,j}}{y_{i,j}} \mu(i,j)^{y_{i,j}} \left(1-\mu(i,j)\right)^{\pi_{i,j}-y_{i,j}}.$$

Then

$$\begin{aligned} U_\mu^F(\beta, P) &= \mathbb{E}^\mu\left[u\left(\beta \cdot Y^P\right)\right] = \\ &= \sum_y u(\beta \cdot y)\, \mu\left(Y^P = y\right) = \\ &= \sum_y u(\beta \cdot y) \prod_{i,j} \binom{\pi_{i,j}}{y_{i,j}} \mu(i,j)^{y_{i,j}} \left(1-\mu(i,j)\right)^{\pi_{i,j}-y_{i,j}}. \end{aligned}$$

Example 3.1. *As an example, we can consider an agent utility function of the type*

$$u(\beta \cdot y) = -e^{-\sum_{i,j} \phi_{i,j}\beta_{i,j}y_{i,j}}, \tag{3.1}$$

where $\phi_{i,j}$ are positive constants. It is easy to prove that, under the hypotheses of Remark 3.2,

$$U_\mu^F(\beta, P) = -\prod_{i,j} \left(1 - \mu(i,j) + \mu(i,j)\, e^{-\phi_{i,j}\beta_{i,j}}\right)^{\pi_{i,j}}. \tag{3.2}$$

In the "degenerate" case, $\phi_{i,j} = k$ for every i and j, then the agent is not interested in the specific class to which each client belongs and her expected utility depends only on the total wealth W.

The company expected utility will depend on $\beta \in \Delta$, on $P \in \mathcal{P}$ and on other variables $\overline{x} = (x_1, x_2, \ldots, x_m) \in \mathcal{D} \subset \mathbb{R}^m$, such as premiums, charges, performances for life or death of the policy owners, that are independent of the relationship between the company and the agent.

Let $U^L : \mathcal{D} \times \Delta \times \mathcal{P} \to \mathbb{R}$ be the company's expected utility function

$$U_\mu^L(\overline{x}, \beta, P) = \mathbb{E}^\mu\left[v\left(\overline{x}, (\mathbf{1} - \beta) \cdot Y^P\right)\right], \tag{3.3}$$

where $\mathbf{1} \in \mathbb{M}$ is the matrix with all entries equal to 1 and $v : \mathcal{D} \times \mathbb{M} \to \mathbb{R}$ is such that $v(\overline{x}, A)$ is an increasing function with respect to each entry of A.

Remark 3.3. *Remark 3.1 holds for the utility function of the company as well.*

Example 3.2. *Similarly as in Example 3.1, we consider a company utility function of the type*

$$v(\overline{x}, (\mathbf{1} - \beta) \cdot y) = -e^{-\sum\limits_{i,j} \psi_{i,j}(1-\beta_{i,j}) y_{i,j}}, \tag{3.4}$$

where $\psi_{i,j}$ are positive constants. It is easy to prove that, under the hypotheses of Remark 3.2,

$$U_{\mu}^{L}(\overline{x}, \beta, P) = -\prod_{i,j} \left(1 - \mu(i,j) + \mu(i,j)\, e^{-\psi_{i,j}(1-\beta_{i,j})}\right)^{\pi_{i,j}}. \tag{3.5}$$

Let $\tilde{\Delta} \subset \Delta$ represent the set of all participating rates the company is actually propense to pay to the agent. Such a subset may be selected, for example, via a preliminary profit test analysis.

The game develops as follows. The company describes to the agent the characteristics of the new policy and proposes a certain $\beta \in \tilde{\Delta}$. The agent analyzes the company proposal and adopts a strategy which maximizes her expected utility function.

First of all, we have to consider how the agent may value the company proposal. On the basis of the portfolio of her clients and the type of policy holders she can reach, the agent will accept to sell the policy for the company if

$$\exists P \in \mathcal{P} : U_{\mu}^{F}(\beta, P) \geq L_A,$$

where L_A is the minimum expected utility value the agent wants to attain.

Therefore, defining, for each β the set $\mathcal{A}_{\mu}(\beta) = \left\{P \in \mathcal{P} : U_{\mu}^{F}(\beta, P) \geq L_A\right\}$, the game is formally the following.

I Level:

$$\max_{\beta \in \tilde{\Delta}} U_{\mu}^{L}(\overline{x}, \beta, P), \tag{3.6}$$

where P, for a fixed β, solves

II Level:

$$\max_{P \in \mathcal{A}_\mu(\beta)} U_\mu^F(\beta, P). \tag{3.7}$$

Remark 3.4. *The distribution μ may depend on $\bar{x}$, but, since $\bar{x}$ is not relevant in the game, this case is not interesting.*

Remark 3.5. *The maximum in (3.7) is surely reached since the expected utility is optimized on a finite set.*

The maximum in (3.6) is reached if we suppose the utility function $v(\bar{x}, A)$ (see (3.3)) to be continuous with respect to A and $\tilde{\Delta}$ to be closed.

3.1. Hierarchical Relationship between the Agent and the Company

In the game model we have proposed, we suppose that the company knows exactly the distribution μ adopted by the agent. This can be true, for example, for the bank-insurance company, but in general the distribution μ is not well-known to the company. However, if the agent has worked for the company in the past, the company can make some hypotheses on μ.

Specifically, we assume that the company knowledge of the agent's behavior is modeled as follows. The company knows the utility function u, but is not able to identify the distribution μ; hence, it will consider a set $\mathcal{M}$ of possible distributions.

Therefore, the utility of the company is given by:

$$\inf_{\mu \in \mathcal{M}} U_\mu^L(\bar{x}, \beta, P).$$

The set $\mathcal{M}$ does not need to satisfy any particular property, since the infimum in the above definition is always finite. In fact $U_\mu^L(\bar{x}, \beta, P) \geq v(\bar{x}, \mathbf{1}, \mathbf{0})$, for every $\mu \in \mathcal{M}$, where $\mathbf{1} \in \mathbb{M}$ is the matrix with all entries equal to 1 and $\mathbf{0} \in \mathbb{M}$ is the matrix with all entries equal to 0.

Then the game becomes:

I Level:

$$\max_{\beta \in \tilde{\Delta}} \inf_{\mu \in \mathcal{M}} U_\mu^L(\bar{x}, \beta, P(\mu)),$$

where $P(\mu)$, for a fixed β, solves

II Level:

$$\max_{P \in \mathcal{A}_\mu(\beta)} U_\mu^F(\beta, P).$$

Remark 3.6. *If the distribution $\hat{\mu}$ considered by the agent to determine her strategy is not in $\mathcal{M}$, the game doesn't work and there is not a hierarchical relationship between the company and the agent. If $\hat{\mu} \in \mathcal{M}$, then the actual utility of the company verifies*

$$U_{\hat{\mu}}^L(\overline{x}, \beta, P(\hat{\mu})) \geq \inf_{\mu \in \mathcal{M}} U_\mu^L(\overline{x}, \beta, P(\mu))$$

and the size of $\mathcal{M}$ is related to the knowledge the company possesses of the behavior of the agent.

4. Agent's Indifference to a Set of Strategies

It may happen that, for some β proposed by the company, there exists a set $O(\beta)$ of strategies P that solve problem (3.7) (this case, in a general setting, has been introduced in subsection 2.2). Obviously each $P \in O(\beta)$ is equivalent for the agent, but in general it is not so for the company.

Example 3.3. *Let us consider a very simple case, with $K = \{1\}$, two income level ranges, $L_A = -\infty$ and an expected utility function of the agent of the type (3.2), that is*

$$U_\mu^F(\beta, P) = -\left(1 - \mu(1,1) + \mu(1,1)\, e^{-\phi_{1,1}\beta_{1,1}}\right)^{\pi_{1,1}} \cdot \\ \cdot \left(1 - \mu(1,2) + \mu(1,2)\, e^{-\phi_{1,2}\beta_{1,2}}\right)^{\pi_{1,2}}.$$

The problem the agent has to solve is, for each β,

$$\max_{0 \leq \pi_{1,1} + \pi_{1,2} \leq N_A} -\left(1 - \mu(1,1) + \mu(1,1)\, e^{-\phi_{1,1}\beta_{1,1}}\right)^{\pi_{1,1}} \cdot \\ \cdot \left(1 - \mu(1,2) + \mu(1,2)\, e^{-\phi_{1,2}\beta_{1,2}}\right)^{\pi_{1,2}}. \tag{4.8}$$

Now, if

$$1-\mu(1,1)+\mu(1,1)\,e^{-\phi_{1,1}\beta_{1,1}}=1-\mu(1,2)+\mu(1,2)\,e^{-\phi_{1,2}\beta_{1,2}}$$

(for example $\mu(1,1)=2\mu(1,2)$*,* $\phi_{1,1}=5$*,* $\phi_{1,2}=1$*,* $\beta_{1,1}=0.02$ *and* $\beta_{1,2}=0.0211$*), then all* $(\pi_{1,1},\pi_{1,2})$ *such that* $\pi_{1,1}+\pi_{1,2}=N_A$ *solve (4.8).*

Referring to Mignanego-Mulinacci ([11]), we model the uncertainty the company has about the agent's strategy choice in $O(\beta)$, by assuming that the company associates, to every β, a probability distribution $\mathbb{P}_\beta$ on $O(\beta)$.

Hence let

$$M_l(\overline{x})=\max_{(\beta,P)\in\tilde{\Delta}\times\mathcal{P}}U_\mu^L(\overline{x},\beta,P)=U_\mu^L(\overline{x},\beta^*,P^*)\,.$$

Even if (β^*,P^*) exists from a mathematical point of view, this is not necessarily true when the game is carried on. In fact, an optimal strategy for the agent is not necessarily optimal for the company.

Then, the company may consider an interval of the type

$$I_\varepsilon=[M_l(\overline{x})-\varepsilon,M_l(\overline{x})]$$

and may announce some $\hat{\beta}$ solving

$$\max_{\beta\in\tilde{\Delta}}\mathbb{P}_\beta\left(U_\mu^L(\overline{x},\beta,P)\in I_\varepsilon\right) \tag{4.9}$$

with the purpose to maximize the probability that a solution of the problem could be as close as possible to $M_l(\overline{x})$.

5. Case Study on a Unit-Linked Policy

Precisely by utilizing the data provided by an insurance company which aimed to create a new life insurance product to reach a specific segment of the market, the authors implemented the Stackelberg game model shown in Section 3 to construct a suitable policy satisfying the actuarial characteristics required by the company.

Let us consider a unit-linked policy with no guarantee. The company pays to the policy holder the value of the units of the fund the policy is linked to, at maturity time T, if the policy owner is alive or, otherwise, in the case of death, a fixed percentage of the value the fund has at the end of the year when the event occurs.

The policy time of maturity is fixed in ten years and there is no possibility to end the contract before. As previously mentioned, the amount in the life case is linked to the performance of an investment fund on the capital market. There are three funds with different risk profiles the policy holder can choose at time 0: weak, medium or aggressive.

The initial premium payable by the policy owner at time 0 is fixed in 10.000 Euros.

The policy has a fixed cost of 60 Euros at the beginning; each year revaluated using a 2% rate. There are also charges for acquisition (5.50% of the net premium) and administration (3.50% of the net premium) a year.

Net yearly premium charges for managing the investment fund are also present.

We are going to work in a simplified framework in which there not exist recursive premiums, guarantees, possibility of switching among different risk profiles or ending the contract before the stipulated date.

The insurance company wants to sell the policy to a particular set of clients consisting of professional men with a high income a year.

We can consider three different workers categories, that is, the set K of all considered jobs (see Section 3) given by $K = \{1,2,3\}$, denoting, respectively, professional men, managers, employees plus pensioners and look at four income classes, numbered by $j = 1,2,3,4$, representing, in Euros, respectively, the income range $[0, 15000]$, the income range $(15000, 32600]$, the income range $(32600, 70000]$ and all the incomes greater than 70000.

We wants to focus on the capability of the model in maximizing as much as possible the sale of this particular policy on the specified target of clients.

In order to reach its goal, the company offers to the agent different rappels for different classes of clients. This way the agent will be forced to sell the product to the client class that maximizes the utility of the insurance company. We will use leader-follower games to determine the optima rappels for the agents looking at the insurance company and the agent as the two parts in a hierarchic game.

Let the follower's expected utility function be, for example, the one described in Example 3.1, depending on the rappel the company will pay and on the number of unit-linked policies sold. She wants to play the game maximizing her expected utility. This way, she may find a set of selling strategies on which she has the same utility level. In this particular case, we associate a probability distribution to the above set according to the knowledge of the company about the past behavior of the agent.

Then we maximize the probability that the expected utility of the insurance company be as close as possible to the global maximum (see subsection 3.2). If the company wants to have a solution closer to the global maximum, the game could be repeated in many time-intervals so let to generate a dynamic game.

4.1. Numerical Results

On the basis of the policy conditions described above, the authors have implemented their model. Here we shall present and discuss our results.

The maximum value of the percentage of the net premium to be given to the agent that the policy can support to guarantee a reasonable income for the company is 3.5%; such a value has been computed by a preliminary profit test analysis on the unit-linked policy with fixed rappels.

The company guarantees, anyway, to the agent a 1% fixed percentage of the net premium for every sold policy independent of the purchaser's type. In addition, it decides to consider a set $\tilde{\Delta}$ of admissible rappels of the type

$$\tilde{\Delta} = \left\{ \begin{pmatrix} 0.001 & 0.003 & \beta_{1,3} & \beta_{1,4} \\ 0 & 0.002 & \beta_{2,3} & \beta_{2,4} \\ 0 & 0 & 0 & 0.2 \end{pmatrix} : \beta_{i,j} \in \{0.02, 0.0225, 0.025\}, i = 1,2, j = 3,4 \right\}.$$

The maximum number of visits the agent makes in the fixed time interval is 8.

In the computations we assumed that the agent's and the company's utilities are (3.2) and (3.5). As far as the utility of the agent is concerned (see Example 3.1) we have supposed $\phi_{i,j} = 1$ for all i and j, assuming that her utility solely depends on her earning. Taking the logarithm, the optimization problems to be solved become linear programming problems.

In the following, we shall use the notations:

- C: matrix whose elements represent the maximum numbers of contacts the agents can have with each class;
- S: matrix containing the selling probabilities, meaning its elements are the $\mu(i,j)$ introduced in Section 3;
- $\hat{W}$: matrix whose elements are the weights associated to each class in the choice of the client the policy is proposed to (this way the distribution $\mathbb{P}_\beta$ of subsection 3.2 is constructed);

- Λ: matrix whose elements are the coefficients $\psi_{i,j}$ considered in the expression of the company's utility (see Example 3.2);
- Q: maximum probability (see (4.9));
- B^{opt}: set of all optimal rappels;
- m: matrix whose elements are the mean numbers of sold policies. Such a mean is computed weighting the mean number of the sold policies for each strategy by the probability that the agent has to choose that strategy, i.e.

$$\sum_{P \in O(\beta)} \mathbb{P}_\beta (P) \sum_{i,j} \pi_{i,j} \mu(i,j).$$

Example 4.1. *Let us consider the case*

$$C = \begin{pmatrix} 0 & 5 & 5 & 5 \\ 6 & 6 & 7 & 6 \\ 5 & 6 & 7 & 8 \end{pmatrix}, \qquad \Lambda = \begin{pmatrix} 0.5 & 0.5 & 1.8 & 2.0 \\ 0.5 & 0.5 & 1.5 & 1.8 \\ 0.5 & 0.5 & 0.5 & 0.5 \end{pmatrix}$$

$$S = \begin{pmatrix} 0.02 & 0.1 & 0.3 & 0.4 \\ 0.05 & 0.1 & 0.3 & 0.4 \\ 0 & 0 & 0.2 & 0.3 \end{pmatrix}, \qquad \hat{W} = \begin{pmatrix} 0.8 & 0.8 & 0.8 & 0.8 \\ 0.2 & 0.2 & 0.2 & 0.2 \\ 0.1 & 0.1 & 0.1 & 0.1 \end{pmatrix}.$$

This way,
$M_l(\overline{x}) = -0.0363951$, *and, posing* $\varepsilon = 0.00007834$, *then* $Q = 0.752941$.

The set of all optima β *for the company is:*

$$B^{opt} = \left\{ \begin{pmatrix} 0.001 & 0.003 & \beta_{13} & 0.02 \\ 0 & 0.002 & \beta_{23} & 0.02 \\ 0 & 0 & 0 & 0.02 \end{pmatrix} : \beta_{i,3} \in \{0.02, 0.0225, 0.025\}, i = 1,2 \right\}$$

while

$$m = \begin{pmatrix} 0 & 0 & 0 & 1.872941176 \\ 0 & 0 & 0 & 1.327058824 \\ 0 & 0 & 0 & 0 \end{pmatrix}.$$

Some of the obtained results, that we shall comment in this example, varying ε *and* $\hat{W}$ *are shown in Appendix 6.1, Case 1.*

In every case we note the existence of a value $\varepsilon = \hat{\varepsilon}$, so that the maximum probability defined in (4.9) bounces between smaller and higher values.

The sensitivity analysis carried out in this study provides an important information to the company. Contrary to what may seem natural at first, the company is able to force the agent to propose the policy to the clients in the classes of managers and professionals with the highest income without necessarily guaranteeing the maximum admissible rappel in case the agent succeeds in selling the insurance plan to a client in such classes. As a matter of fact, analyzing the matrix that contains the mean number of the sold policies, we can notice that the selling strategies of the agent are concentrated only on the classes of managers and professionals with the highest income.

Such a result is strictly related to the composition of the selling probabilities matrix S in which the highest values are exactly in these classes.

¿From an economical point of view, this result is a consequence of the company's capability of creating a personalized policy that covers, as much as possible, the needs of the clients it wants to reach. This way, it becomes easier for the agent to sell the policy to the clients in the classes the company aims at.

Thus the company, if it is interested in reducing the costs linked to the rappels, must be able to construct a policy having a high probability to be appealing to the classes it intends to reach.

In these cases, the optimal solutions given by the model show this evidence: the rappels the company proposes to the agent for selling the policy to the other classes do not appear relevant.

Such a remark is based on the particular composition of the clients portfolio of the agent, as shown below.

Example 4.2. *Let us consider the case*

$$C = \begin{pmatrix} 0 & 3 & 3 & 3 \\ 2 & 3 & 4 & 3 \\ 1 & 2 & 3 & 3 \end{pmatrix}, \qquad \Lambda = \begin{pmatrix} 0.5 & 0.5 & 1.8 & 2.0 \\ 0.5 & 0.5 & 1.5 & 1.8 \\ 0.5 & 0.5 & 0.5 & 0.5 \end{pmatrix}$$

$$S = \begin{pmatrix} 0.02 & 0.1 & 0.7 & 0.8 \\ 0.05 & 0.1 & 0.5 & 0.6 \\ 0 & 0 & 0.2 & 0.3 \end{pmatrix}, \qquad \hat{W} = \begin{pmatrix} 0.5 & 0.5 & 0.5 & 0.5 \\ 0.5 & 0.5 & 0.5 & 0.5 \\ 0.5 & 0.5 & 0.5 & 0.5 \end{pmatrix} (uniform).$$

This way,

$$M_l(\overline{x}) = -0.00057245, \text{ and, if } \varepsilon = 9.9942 \cdot 10^{-8}, \text{ then } Q = 1.0$$

and the set of all optimal β *for the company is:*

$$B^{opt} = \left\{ \begin{pmatrix} 0.001 & 0.003 & 0.02 & 0.02 \\ 0 & 0.002 & \beta_{2,3} & 0.02 \\ 0 & 0 & 0 & 0.02 \end{pmatrix} : \beta_{2,3} \in \{0.02, 0.0225\} \right\}$$

while

$$m = \begin{pmatrix} 0 & 0 & 2.1 & 2.4 \\ 0 & 0 & 0 & 1.2 \\ 0 & 0 & 0 & 0 \end{pmatrix}.$$

In this case the company reaches his optimal maximum $M_l(\overline{x}) = -0.00057245$.

Some of the obtained results, that we shall comment in this example, varying ε *and S are shown in Appendix 6.2, Case 2.*

Compared to the results discussed in the previous Example, a different composition of the matrix of the clients portfolio of the agent has been considered. The total number of potential buyers in the classes of managers and professionals with the highest income levels does not cover the considered number of possible visits.

Thus the agent is forced to also contact clients of other classes.

The results show that the selling probabilities determine strategies for the agent that concentrates on the classes with such a higher probability. In fact, if such maximum probabilities are equal, then also the mean number of sold policies are the same.

It is important to notice that, in any case, we have the minimum rappel value on the classes with a higher insurance policy acquisition probability; these coincide with that segment the company intends to reach.

Remark 4.1.

Analyzing the sensitivity of the results varying the coefficients of the company's utility function, we observe that the remarks mentioned for the previous cases remain valid, that is the most important contribution is given again by the probability the policy has to be sold to the different classes. Clearly, the combination of the composition of the matrix S and of the composition of the matrix Λ *influences the optimal rappels.*

Remark 4.2.

Finally, observing the sensibility of the results varying the probability distribution that the company supposes on the choice of the agent among his optimal strategies, the analysis presented for the previous cases remain valid.

Furthermore, we may notice that, if the company has consistent information on the agent, it can build a product having such characteristics that put her in the position of selling the policy to specific classes of clients without paying her the maximum admissible rappel.

5. Conclusion

The proposed game model gives interesting results compared to the possible strategy of the company in creating an insurance product for a specific target of clients.

If the proposed policy is constructed in such a way to appeal to the class of clients the company wants to reach, she will be able to propose to the agent a rappel even smaller than the maximum possible even continuing in having a reasonable utility.

Such a strategy is also well accepted by the agent who, as shown by the results, prefers to propose the policy to the clients with higher probability to buy it despite having a lower rappel.

6. Appendix

In this Appendix we present a selection of the obtained results that seem to be more significant to understand the conclusions.

6.1. Case 1

Results A.

$$\hat{W} = \begin{pmatrix} 0.8 & 0.8 & 0.8 & 0.8 \\ 0.2 & 0.2 & 0.2 & 0.2 \\ 0.1 & 0.1 & 0.1 & 0.1 \end{pmatrix},$$

$$M_l\left(\overline{x}\right) = -0.0363951,\, \varepsilon = 0.00007840,\, Q = 1.0.$$

The set of all optimal β for the company is:

$$B^{opt} = \left\{ \begin{pmatrix} 0.001 & 0.003 & \beta_{1,3} & 0.0225 \\ 0 & 0.002 & \beta_{2,3} & 0.02 \\ 0 & 0 & 0 & 0.02 \end{pmatrix} : \beta_{i,3} \in \{0.02, 0.0225, 0.025\}, i = 1,2 \right\};$$

$$m = \begin{pmatrix} 0 & 0 & 0 & 2.0 \\ 0 & 0 & 0 & 1.2 \\ 0 & 0 & 0 & 0 \end{pmatrix}.$$

Results B.

$$\hat{W} = \begin{pmatrix} 0.5 & 0.5 & 0.5 & 0.5 \\ 0.5 & 0.5 & 0.5 & 0.5 \\ 0.5 & 0.5 & 0.5 & 0.5 \end{pmatrix} \text{ (uniform)},$$

$$M_l(\overline{x}) = -0.0363951, \varepsilon = 0.00007834, Q = 0.25.$$

The set of all optimal β for the company is:

$$B^{opt} = \left\{ \begin{pmatrix} 0.001 & 0.003 & \beta_{1,3} & 0.02 \\ 0 & 0.002 & \beta_{2,3} & 0.02 \\ 0 & 0 & 0 & 0.02 \end{pmatrix} : \beta_{i,3} \in \{0.02, 0.0225, 0.025\}, i = 1,2 \right\};$$

$$m = \begin{pmatrix} 0 & 0 & 0 & 1.4 \\ 0 & 0 & 0 & 1.8 \\ 0 & 0 & 0 & 0 \end{pmatrix}.$$

Results C.

$$\hat{W} = \begin{pmatrix} 0.5 & 0.5 & 0.5 & 0.5 \\ 0.5 & 0.5 & 0.5 & 0.5 \\ 0.5 & 0.5 & 0.5 & 0.5 \end{pmatrix} \text{ (uniform)},$$

$$M_l(\overline{x}) = -0.0363951, \varepsilon = 0.00007840, Q = 1.0.$$

The set of all optimal β for the company is:

$$B^{opt} = \left\{ \begin{pmatrix} 0.001 & 0.003 & 0.02 & 0.0225 \\ 0 & 0.002 & 0.025 & 0.02 \\ 0 & 0 & 0 & 0.02 \end{pmatrix} : \beta_{i,3} \in \{0.02, 0.0225, 0.025\}, i = 1,2 \right\};$$

$$m = \begin{pmatrix} 0 & 0 & 0 & 2.0 \\ 0 & 0 & 0 & 1.2 \\ 0 & 0 & 0 & 0 \end{pmatrix}.$$

6.2. Case 2

Results A.

$$S = \begin{pmatrix} 0.02 & 0.1 & 0.3 & 0.4 \\ 0.05 & 0.1 & 0.3 & 0.4 \\ 0 & 0 & 0.2 & 0.3 \end{pmatrix},$$

$$M_l(\overline{x}) = -0.0477015,\ \varepsilon = 0.00005951,\ Q = 1.0.$$

The set of all optimal β for the company is:

$$B^{opt} = \left\{ \begin{pmatrix} 0.001 & 0.003 & 0.025 & 0.02 \\ 0 & 0.002 & \beta_{2,3} & 0.02 \\ 0 & 0 & 0 & 0.02 \end{pmatrix} : \beta_{2,3} \in \{0.02, 0.0225\} \right\};$$

$$m = \begin{pmatrix} 0 & 0 & 0.6 & 1.2 \\ 0 & 0 & 0 & 1.2 \\ 0 & 0 & 0 & 0 \end{pmatrix}.$$

Results B.

$$S = \begin{pmatrix} 0.02 & 0.1 & 0.3 & 0.4 \\ 0.05 & 0.1 & 0.3 & 0.4 \\ 0 & 0 & 0.2 & 0.3 \end{pmatrix},$$

$$M_l(\overline{x}) = -0.0477015,\ \varepsilon = 0.00005601,\ Q = 0.\overline{3}.$$

The set of all optimal β for the company is:

$$B^{opt} = \left\{ \begin{pmatrix} 0.001 & 0.003 & 0.0225 & 0.02 \\ 0 & 0.002 & 0.0225 & 0.02 \\ 0 & 0 & 0 & 0.02 \end{pmatrix} \right\};$$

$$m = \begin{pmatrix} 0 & 0 & 0.3 & 1.2 \\ 0 & 0 & 0.3 & 1.2 \\ 0 & 0 & 0 & 0 \end{pmatrix}.$$

References

[1] R.J. Aumann, *Game Theory*, in The new Palgrave Dictionary of Economics, (by: J.Eatwell, M. Milgate e P. Newman), Macmillan, London, 1987, 460–482.

[2] T. Başar, G. Olsder, *Dynamic Noncooperative Game Theory*, Academic Press, New York, 1995.

[3] J. Bard, Y. Ishizuka, K. Shimizu, *Nondifferentiable and Two level programming*, Kluwer, 1997.

[4] I. Colivicchi, *Le 'unit-linked': proposta di sintesi delle principali caratteristiche attuariali sul mercato italiano*, Contributi di ricerca in Econometria e Matematica, *Quaderno dell'Istituto di Econometria e Matematica per le applicazioni Economiche, Finanziarie ed Attuariali dell'Universitá Cattolica di Milano* **29** (2003).

[5] P.C. Fishburn, *Utility Theory for Decision Making*, Robert E. Krieger Publishing Co., Huntington, New York, 1979.

[6] D. Fudenberg, J. Tirole, *Game Theory*, MIT Press, Cambridge (Massachusetts), 1991.

[7] J.E. Ingersoll, *Theory of Financial Decision Making*, Rowman & Littlefield Publishers, Savage (Maryland), 1987.

[8] G. Leitmann, On generalized Stackelberg Strategies, *J. Opt. Th. And Appl.* **26** (1978), 637–643.

[9] L. Mallozzi, J. Morgan, *Hierarchical system with weighted reaction set*, Nonlinear optimization, Dipillo and Giannessi Eds., Pitman, (1996), 271–283.

[10] R.B. Myerson, *Game Theory: Analysis and Conflict*, Harvard University Press, Cambridge (Massachusetts), 1991.

[11] F. Mignanego, S. Mulinacci, *Credibility and stochastic leader-follower games*, Contributi di ricerca in Econometria e Matematica,*Quaderno dell'Istituto di Econometria e Matematica per le applicazioni Economiche, Finanziarie ed Attuariali dell'Universitá Cattolica di Milano* **68** (2003).

[12] A. Rubinstein, O. Rubinstein, *A Course on Game Theory*, MIT Press, Cambridge (Massachusetts), 1994.

[13] L. Vannucci, On estimation of automobile insurance fraudolence degee, *Giornale Istituto Italiano degli Attuari* **LXVI** (2003), 83–99.

INDEX

A

B

C

D

E

F

G

H

I

J

K

L

M

N

O

P

Q

R

S

T

U

V

W

Y